A Practical Approach
to Conditions of Contract
for Civil Engineering Works

A Practical Approach to Conditions of Contract for Civil Engineering Works

Ir David Y. K. LEUNG

BSc, M(Eng), Dip Arb, FCIArb, MHKIE

in association with
LEUNG Chung Kee

香港大學出版社
HONG KONG UNIVERSITY PRESS

Hong Kong University Press
14/F Hing Wai Centre
7 Tin Wan Praya Road
Aberdeen
Hong Kong

© Hong Kong University Press 2010

ISBN 978-962-209-178-8

Secure On-line Ordering
http://www.hkupress.org

British Library Cataloguing-in-Publication Data
A catalogue record for this book is available from the British Library.

Printed and bound by Condor Production Ltd., Hong Kong, China

Contents

Foreword

When David told me about his project on writing a book about the practical use of the current edition of the *General Conditions of Contract for Civil Engineering Works*, I immediately congratulated him on coming up with such a great idea. It is beyond my doubt that David is the right person to do this.

The HKSAR Government has adopted the General Conditions of Contract for all its civil engineering contracts. The practitioners in the engineering sector have been longing for a reference on the subject. Some of them have resorted to references written for other countries though the relevance of some of their contents is not guaranteed. David's book will therefore provide the engineering professionals and other related professionals with a practical guide to the subject with specific reference to Hong Kong.

David has the right credentials to write this book. He is a qualified engineer, a Fellow of the Chartered Institute of Arbitrators, as well as a holder of the Postgraduate Certificate in Laws. He has been working in the engineering profession for 30 years, during which he has acquired solid experience in administering major contracts. His more than 12 years of teaching experience on the subject also give him a good idea on the types of contents to be included in a comprehensive reference.

The publication of David's book coincides with one of the most exciting periods of the construction sector in Hong Kong. Over the next few years, the HKSAR Government will implement a major infrastructure development programme, including the 10 major infrastructure projects highlighted in the 2007 Policy Address. The timely release of his book will surely be great news for engineers and related professionals, who will finally be able to have a handbook on the General Conditions of Contract that is tailor-made for Hong Kong.

I am confident that David's endeavours in this book will surely meet with success.

Ir Dr the Honourable Raymond HO Chung-tai
Member of the Legislative Council (Engineering Functional Constituency)
Hong Kong Deputy to the 11th National People's Congress

Preface

This book provides a comprehensive commentary and guidance to readers on the current edition (i.e., the 1999 Edition) of the *General Conditions of Contract for Civil Engineering Works*, which the Government of the HKSAR uses for all its civil engineering contracts. This is the first book of its kind ever written in Hong Kong.

Out of the 90 clauses in the General Conditions of Contract, 46 clauses are chosen. These represent the majority of frequently used and commonly discussed clauses, as well as the most topical clauses in seminars and lectures.

Students, young engineers and technical persons working on construction sites often find it difficult to understand the General Conditions of Contract because they are not written in plain English and the sentences are unduly long and complicated. On the other hand, the UK ICE Conditions and the international FIDIC Conditions are written in much user-friendly English. The listing of equivalent clauses together enables the readers to understand the meaning of the clauses from a different context. For those readers who find it easier to read in Chinese, the translation will help them to compare with and understand the original English text. To read between lines and understand the underlying meaning of the clauses, the reader should turn to the sections "Commentary" and the "Analysis and application," which are written in plain and user-friendly English.

For practitioners, the book can be used as a comprehensive reference. Apart from analyzing each clause with due consideration of construction practice in Hong Kong, the book also refers to relevant legal cases and citations from authoritative text books, such as those written by Mr. Max Abrahamson and Mr. Duncan Wallace.

The comparison with the UK ICE Conditions of Contract and the international FIDIC Conditions of Contract is interesting and points to the possibility of improving the HKSAR General Conditions of Contract to be compatible with the most current edition of the English and international standards in term of readability, procedural aspects and equitable sharing of risk between the Contractor and the Employer.

Ir David Y. K. LEUNG

Acknowledgements

First of all, I would like to thank Ir Dr the Honorable Raymond Ho Chung-tai, who spent his most valuable time during his 2008 Legislative Council Election period writing the Foreword for me. I also congratulate him for being elected again as a Legislative Council member in the 2008 Legislative Council Election.

I am particularly indebted to the Government of the Hong Kong Special Administrative Region for permission to write this book. I am also indebted to the Federation International Des Ingenieurs-Conseils (FIDIC) for permission to reproduce the clauses from the *FIDIC Conditions for Construction* (1999 Edition). Likewise, I am indebted to the Institution of Civil Engineers, Association of Consulting Engineers and Civil Engineering Contractors Associations, as sponsoring authorities, for permission to reproduce the clauses from the *Institution of Civil Engineering Conditions of Contract for Construction* (7th Edition).

I am particularly thankful to my brother, Mr. C. K. Leung for his help in producing this book and for his efforts in the Chinese translation. The completion of this book would not have been possible without his contribution and encouragement. I also gratefully acknowledge the assistance of my good friend, Mr. Paul Chong, in the Chinese translation.

To my sister-in-law, Eva Wong, I gratefully acknowledge her dedication and patience in typing and finalizing the layout, and to my loving daughter, Bianca, who designed the cover of this book.

Last but not least, I also wish to express my gratitude to my beloved wife, Christine, who affectionately took care of the well-being of the whole production team in the long period during which this book was compiled.

Ir David Y. K. LEUNG

Introduction

This book describes in detail the applications of 46 most important and essential clauses out of the 90 clauses in the current edition (1999 Edition) of the *General Conditions of Contract for Civil Engineering Works* (the "HKSAR Conditions") published by the Government of the Hong Kong Special Administrative Region (HKSAR). These clauses are compared with the corresponding clauses in the current edition (7th Edition) of the UK *Institution of Civil Engineering Conditions of Contract* (the "ICE Conditions") and the current edition (1999 Edition) of the international *FIDIC Conditions for Construction* (the "FIDIC Conditions") (referred to as the "HKSAR Clauses," "ICE Clauses" and "FIDIC Clauses"). For each clause there is a commentary which describes the general application of the clause, highlighting any special areas or problems arising from its past usage. The sentences and phrases of each clause/sub-clause are also meticulously analyzed, with references to academic and legal authorities as appropriate. The commentary draws experience from the author's profession as a practising engineer and contract adviser.

Each clause of the HKSAR Conditions is followed by its Chinese translation to assist understanding, and the corresponding clauses in the ICE Conditions and the FIDIC Conditions for easy comparison. The author's analysis and commentary on the clause, and its comparison with the ICE Clauses and the FIDIC Clauses are also included. The text is interspersed with photographs of major projects, each with a relevant question to further illustrate the application and test the reader's understanding of the HKSAR Conditions. Suggested answers to the questions are given in the Appendixes.

The Chinese translation of the clauses of the HKSAR Conditions is not the official version and is for reference only.

The HKSAR Conditions:

Definitions and Interpretation

(1) In the Contract the following words and expressions shall have the meaning hereby assigned to them except when the context otherwise requires:

"Contract" means the Articles of Agreement, the Tender and the acceptance thereof by the Employer (including such further agreed documents as may be expressly referred to in or by the same: Drawings, General Conditions of Contract, Special Conditions of Contract (if any), Specification and priced Bills of Quantities.

"Contractor" means the person, firm or company whose Tender has been accepted by the Employer and includes the Contractor's personal representatives, successors and permitted assigns.

"Drawings" means the drawings referred to in the Specification or Bills of Quantities and any modification of such drawings approved in writing by the Engineer and such other drawings as may from time to time be furnished in writing or approved in writing by the Engineer.

"Employer" means the Government of the Hong Kong Special Administrative Region.

"Engineer" means the person, company or firm appointed from time to time by the Employer and notified in writing to the Contractor to act as the Engineer for the purposes of the Contract. The person appointed may be described by name or as the holder for the time being of a public office.

"Engineer's Representative" means any person or persons appointed from time to time by the Engineer and notified in writing to the Contractor to perform the duties set forth in Clause 2(2). The person appointed may be described by name or as the holder for the time being of a public office.

Corresponding clauses in the ICE Conditions:

Clause 1

(1) In the Contract (as hereinafter defined) the following words and expressions shall have the meanings hereby assigned to them except where the context otherwise requires.

(a) "Employer" means the person or persons firm company or other body named in the Appendix to the Form of Tender and includes the Employer's personal representatives successors and permitted assignees.

(b) "Contractor" means the person or persons firm or company to whom the Contract has been awarded by the Employer and includes the Contractor's personal representatives successors and permitted assignees.

(c) "Engineer" means the person firm or company appointed by the Employer to act as Engineer for the purposes of the Contract and named in the Appendix to the Form of Tender or any other person firm or company so appointed from time to time by the Employer and notified in writing as such to the Contractor.

(d) "Engineer's Representative" means a person notified as such from time to time by the Engineer under Clause 2(3)(a).

(e) "Contract" means the Conditions of Contract Specification Drawings Bill of Quantities the Form of Tender the written acceptance thereof and the Form of Agreement (if completed).

(g) "Drawings" means the drawings referred to in the Specification and any modification of such drawings approved in writing by the Engineer and such other drawings as may from time to time be furnished by or approved in writing by the Engineer.

Corresponding clauses in the FIDIC Conditions:

Clause 1.1

In the Conditions of Contract ("these Conditions"), which include Particular Conditions, the following words and expressions shall have the meanings stated. Words indicating persons or parties include corporations and other legal entities, except where the context requires otherwise.

1.1.1.1 "Contract" means the Contract Agreement, the Letter of Acceptance, the Letter of Tender, these Conditions, the Specification, the Drawings, the Schedules, and the further documents (if any) which are listed in the Contract Agreement or in the Letter of Acceptance.

1.1.1.6 "Drawings" means the drawings of the Works, as included in the Contract, and any additional and modified drawings issued by (or on behalf of) the Employer in accordance with the Contract.

1.1.2.2 "Employer" means the person named as employer in the Appendix to Tender and the legal successor in title to this person.

1.1.2.3 "Contractor" means the person(s) named as contractor in the Letter of Tender accepted by the Employer and the legal successors in title to this person(s).

1.1.2.4 "Engineer" means the person appointed by the Employer to act as the Engineer for the purposes of the Contract and named in the Appendix to Tender, or other person appointed from time to time by the Employer and notified to the Contractor under Sub-Clause 3.4 [*Replacement of the Engineer*].

條款 1

香港特別行政區土木工程合同一般條件:

定義及解釋

(1) 在這「合同」內,除內文另有需要,下列文字和詞句具有如下文所給予的含意:

「合同」指「協議書」、「投標」及獲僱主接納該「投標」的文件 (包括其載有或在其內文明文提及的其他議訂文件)、「圖則」、「合同一般條件」、「合同特別條件」(如有此文件的話)、「規格」及已標價的「工程量清單」。

「承建商」指其「投標」獲得僱主接納的人、商號或公司,及包括其遺產代理人、承繼人及得到允許的承讓人。

「圖則」指「規格」或「工程量清單」內所提及的圖則及任何獲「工程師」書面核准對此等圖則所作的修訂,及「工程師」可不時以書面供給或經「工程師」書面核准的其他圖則。

「僱主」指香港特別行政區政府。

「工程師」指僱主可不時委任並用書面通知「承建商」作為有關本「合同」事宜擔任工程師的人、公司或商號。對獲委任的人可以他的姓名稱呼,亦可以他當時擔任的職位稱呼。

「工程師代表」指「工程師」可不時委任並用書面通知「承建商」作為履行條款2(2)所列的職責的人。對獲委任的人可以他姓名稱呼,亦可以他當時擔任的職位稱呼。

Commentary

For this definition Clause 1, there is no attempt to analyze each definition, but a comparison of the clause with the corresponding ones of the ICE Conditions and the FIDIC Conditions is made.

Comparison with the ICE Conditions and the FIDIC Conditions

Term in HKSAR Conditions	Points to note in definition and interpretation		
	HKSAR Conditions	ICE Conditions	FIDIC Conditions
Contract	Meaning similar		
Contractor	Meaning similar		
Drawings	Meaning similar		
Employer	The Government of the HKSAR	A person, firm or company	
Engineer	Named in the Letter of Acceptance.	Named in the Appendix to the Form of Tender.	
Engineer's Representative	Meaning similar		

The HKSAR Conditions:

Definitions and Interpretation

(continued)

(1) "Constructional Plant" means all appliances or things of whatsoever nature required for the execution of the Works but does not include materials or other things intended to form or forming part of the permanent work or vehicles engaged in transporting any personnel, Constructional Plant, materials or other things to or from the Site.

"Contingency Sum" means the sum provided for work or expenditure which cannot be foreseen at the time the tender documents are issued which sum may include provision for work to be executed or for materials or services to be supplied by a Nominated Sub-contractor.

"Contract Sum" means the total of the priced Bills of Quantities at the date of acceptance of the Tender for the Works.

"Cost" means expenditure reasonably incurred including overheads whether on or off the Site and depreciation in value of Constructional Plant owned by the Contractor but excluding profit.

"Final Contract Sum" means the sum to be ascertained and paid in accordance with the provisions hereinafter contained for the execution of the Works in accordance with the Contract.

"Nominated Sub-contractor" means and includes all specialists, merchants, tradesmen and the like executing any part of the Works or supplying any materials or services for the Works who shall have been or shall be nominated by the Employer and employed by the Contractor.

"Prime Cost Sum" means the sum provided for work to be executed or for materials or services to be supplied by a Nominated Sub-contractor; such sum shall be the estimated net price to be paid for such work executed or for materials or services supplied by a Nominated Sub-contractor, after deducting any trade or other discount.

"Provisional Sum" means a sum provided for work or expenditure which has not been quantified or detailed at the time the tender documents are issued which sum may include provision for work to be executed or for materials or services supplied by a Nominated Sub-contractor.

"Retention Money" means the sum retained by the Employer as retention money in accordance with the Contract.

Corresponding clauses in the ICE Conditions:

Clause 1 (*continued*)

(1) (i) "Tender Total" means the total of the Bill of Quantities at the date of award of the Contract or in the absence of a Bill of Quantities the agreed estimated total value of the Works at that date.

(j) "Contract Price" means the total of the Bill of Quantities at the date of award of the Contract or in the absence of a Bill of Quantities the agreed estimated total value of the Works at that date.

(k) "Prime Cost[1] (PC) Item" means an item in the Contract which contains (either wholly or in part) a sum referred to as Prime Cost (PC) which will be used for the carrying out of work or the supply of goods materials or services for the Works.

(l) "Provisional Sum" means a sum included and so designated in the Contract as a specific contingency for the carrying out of work or the supply of goods materials or services which may be used in whole or in part or not at all at the direction and discretion of the Engineer.

(m) "Nominated Sub-contractor" means any merchant tradesman specialist or other person firm or company nominated in accordance with the Contract to be employed by the Contractor for the carrying out of work or supply of goods materials or services for which a Prime Cost or a Provisional Sum has been included in the Contract.

(w) "Contractor's equipment" means all appliances or things of whatsoever nature required in or about the construction and completion of the Works but does not include materials or other things intended to form or forming part of the Permanent Works.

Corresponding clauses in the FIDIC Conditions:

Clause 1.1 (*continued*)

1.1.4.2 "Contract Price" means the price defined in Sub-Clause 14.1 [*The Contract Price*], and includes adjustments in accordance with the Contract.

1.1.4.3 "Cost" means all expenditure reasonably incurred (or to be incurred) by the Contractor, whether on or off the Site, including overhead and similar charges, but does not include profit.

1.1.4.10 "Provisional Sum" means a sum (if any) which is specified in the Contract as a provisional sum, for the execution of any part of the Works or for the supply of Plant, Materials or services under Sub-Clause 13.5 [*Provisional Sums*]

1.1.4.11 "Retention Money" means the accumulated retention money which the Employer retains under Sub-Clause 14.3 [*Application for Interim Payment Certificates*] and pays under Sub-Clause 14.9 [*Payment of Retention Money*].

1.1.5.1 "Contractor's Equipment" means all apparatus, machinery, vehicles and other things required for the execution and completion of the Works and the remedying of any defects. However, Contractor's Equipment excludes Temporary Works, Employer's Equipment (if any), Plant, Materials and any other things intended to form or forming part of the Permanent Works.

條款 1

香港特別行政區土木工程合同一般條件：

定義及解釋

（續）

(1) 「施工設備」指實施「本工程」所需要的一切器械或任何物件，無論該等器械或物件屬何等性質，但不包括預定或已經構成永久工程一部份的物料或其他物件，也不包括用以運載任何人員、施工設備、物料或其他物件往返工地的車輛。

「應變金額」指在發出投標文件時未能預測到的工程或支出而預留的金額，此金額可包括為「指定次承判商」實施的工程或供應的物料或提供的服務而預留的金額。

「合同款項」指僱主接納「本工程」「投標」當日已標價的「工程量清單」的總值。

「費用」指合理的支出，包括工地內或工地外的經常開支及「承建商」所擁有的「施工設備」的折舊，但不包括利潤。

「最終合同金額」指按照「合同」實施「本工程」而根據以下條文確定及支付的金額。

「指定次承判商」指及包括所有已是或將由「僱主」指定及得到「承建商」僱用以實施「本工程」任何部分或為「本工程」供應任何物料或提供任何服務的專門人員、商人、技術人員等人。

「基本成本金額」指為「指定次承判商」實施的工程或供應的物料或提供的服務而預留的金額，此金額是在扣除任何商業折扣或其他折扣後為「指定次承判商」實施的工程或供應的物料或提供的服務而估計須支付的淨價格。

「備用金額」指發出投標文件時還未計算數量或詳細列出的工程或支出而預留的金額，此金額可包括為「指定次承判商」實施的工程或供應的物料或提供的服務而預留的金額。

「保留金」指「僱主」按照「合同」保留作為保留金的金額。

Comparison with the ICE Conditions and the FIDIC Conditions

Term in HKSAR Conditions	Points to note in definition and interpretation		
	HKSAR Conditions	ICE Conditions	FIDIC Conditions
Constructional Plant	Includes Temporary Works. Also includes the Engineer's and the Contractor's transport vehicles.	The term "Contractor's Equipment" is used.	
		—	Excludes Temporary Works.
Contingency Sum	—	No such term.	
Contract Sum	—	The term "Contract Price" is used.	
Final Contract Sum	—	No such term.	
Cost	Meaning similar		
Provisional Sum	Meaning similar		
Retention Money	—	No such term is used; the word "retention" is used in its general sense.	
Nominated Sub-contractor	Meaning similar		

The HKSAR Conditions:

Definitions and Interpretation

(continued)

(1)　"General Holiday" means every Sunday and other day which is a general holiday by virtue of the *General Holidays Ordinance* (Cap. 149).

"Hong Kong" means the Hong Kong Special Administrative Region.

"Maintenance Period" means the maintenance period named in the Appendix to the Form of Tender commencing on the day following the date of completion of the Works or any Section or part thereof certified by the Engineer in accordance with Clause 53.

"Portion" means a part of the Site separately identified in the Contract.

"Section" means a part of the Works for which a separate time for completion is identified in the Contract.

"Site" means the lands and other places including the sea under, over, on, in or through which the Works are to be constructed and any other lands or places provided by the Employer for the purpose of the execution of the Works together with such other places as may be subsequently agreed in writing by the Engineer as forming part of the Site.

"Specialist Contractor" means any contractor employed by the Employer to execute Specialist Works.

"Specialist Works" means any work separately identified in the Contract and connected with or ancillary to the Works which may from time to time be carried out on the Site by a Specialist Contractor.

"Specification" means the specifications referred to in the Contract and any modification thereof or addition thereto as may from time to time be furnished in writing or approved in writing by the Engineer.

"Temporary Works" means all temporary work of every kind required for the construction, completion and maintenance of the Works.

"Tender" means the Contractor's tender for the Contract.

"Works" means the work or services including work or services to be carried out by Nominated Sub-contractors to be constructed, completed, maintained and/or supplied in accordance with the Contract and includes Temporary Works.

Corresponding clauses in the ICE Conditions:

Clause 1 *(continued)*

(1)　(f)　"Specification" means the specification referred to in the Form of Tender and any modification thereof or addition thereto as may from time to time be furnished or approved in writing by the Engineer.

(n)　"Permanent Works" means the permanent works to be constructed and completed in accordance with the Contract.

(o)　"Temporary Works" means all temporary works of every kind required in or about the construction and completion of the Works.

(p)　"Works" means the Permanent Works together with the Temporary Works.

(s)　"Defects Correction Period" means that period stated in the Appendix to the Form of Tender calculated from the date on which the Contractor becomes entitled to a Certificate of Substantial Completion for the Works or any Section or part thereof.

(u)　"Section" means a part of the Works separately identified in the Appendix to the Form of Tender.

(v)　"Site" means the lands and other places on under in or through which the Works are to be constructed and any other lands or places provided by the Employer for the purposes of the Contract together with such other places as may be designated in the Contract or subsequently agreed by the Engineer as forming part of the Site.

Corresponding clauses in the FIDIC Conditions:

Clause 1.1 *(continued)*

1.1.1.5　"Specification" means the document entitled specification, as included in the Contract, and any additions and modifications to the specification in accordance with the Contract. Such document specifies the Works.

1.1.3.7　"Defects Notification Period" means the period for notifying defects in the Works or a Section (as the case may be) under Sub-Clause 11.1 [*Completion of*

Outstanding Work and Remedying Defects], as stated in the Appendix to Tender (with any extension under Sub-Clause 11.3 [*Extension of Defects Notification Period*]), calculated from the date on which the Works or Section is completed as certified under Sub-Clause 10.1 [*Taking Over of the Works and Sections*].

1.1.5.4　"Permanent Works" means the permanent works to be executed by the Contractor under the Contract.

1.1.5.6 "Section" means a part of the Works specified in the Appendix to Tender as a Section (if any).

1.1.5.7 "Temporary Works" means all temporary works of every kind (other than Contractor's Equipment) required on Site for the execution and completion of the Permanent Works and the remedying of any defects.

1.1.5.8 "Works" means the Permanent Works and the Temporary Works, or either of them as appropriate.

1.1.6.7 "Site" means the places where the Permanent Works are to be executed and to which Plant and Materials are to be delivered, and any other places as may be specified in the Contract as forming part of the Site.

條款 1

香港特別行政區土木工程合同一般條件：

定義及解釋

（續）

(1) 「公眾假期」指每個星期日及根據公眾假期條例（第149章）定為公眾假期的其他日子。

「香港」指香港特別行政區。

「保養期」指在「投標表格」的「附件」內所指的保養期，由「工程師」根據第53條核實為「本工程」或任何工段或其部分的竣工日期的翌日起計。

「分區」指在「合同」內另分別指定的部分工地。

「工段」指「本工程」的某一部分，而「合同」為此部分分別訂定一個竣工限期。

「工地」指將在其下面、上面、表面、內面或貫穿其中來建造「本工程」的土地及其他地方，包括海在內，亦指「僱主」為「本工程」提供的任何其他土地或地方，以及工程師可於日後以書面同意成為工地一部分的其他地方。

「專門承建商」指任何由「僱主」僱用以實施「專門工程」的承建商。

「專門工程」指在「合同」內分別指定以及與「本工程」有關或附帶於「本工程」而不時由「專門承建商」在「工地」進行的工程。

「規格」指在「合同」內提及的規格，以及「工程師」不時對此等規格以書面提出或核准的任何修訂或增補。

「臨時工程」指為「本工程」的實施、完成及保養所需要的所有各類臨時工程。

「投標」指承建商為「合同」而作出的投標。

「本工程」指按照「合同」而實施、完成、保養及/或提供的工程或服務（包括由「指定次承判商」施行的工程或服務），亦包括「臨時工程」。

Comparison with the ICE Conditions and the FIDIC Conditions

Term in HKSAR Conditions	Points to note in definition and interpretation		
	HKSAR Conditions	ICE Conditions	FIDIC Conditions
Maintenance Period	—	The term "Defects Correction Period" is used.	The term "Defects Notification Period" is used; appears to confine the notification of defects to that period.
Site	Meaning similar; includes places subsequently agreed by the Engineer as forming part of the Site and these places can include those provided by the Contractor such as casting yards and workshops not in the vicinity of the Works.		Excludes such places as provided by the Contractor.
Specialist Contractor	Employed by the Employer, not the Contractor.	No such term	
Specification	Meaning similar		
Temporary Works	Includes Temporary Works for the maintenance of the construction.	Includes Temporary Works up to the completion of the Works.	Includes Temporary Works up to the completion of the Permanent Works but specifically excludes Contractor's Equipment.
Works	Includes works (consisting of both Temporary Works and permanent works) or services.	Includes Permanent Works and Temporary Works but does not explicitly include services.	

Clause 1

Definitions and Interpretation

(continued)

(2) Words importing the singular only also include the plural and vice versa where the context requires.

(3) The index and marginal notes or headings in the General Conditions of Contract, Special Conditions of Contract (if any), and the Specification shall not be deemed to be part thereof or be taken into consideration in the interpretation or construction thereof.

(4) (a) Unless otherwise provided, all payments shall be made in Hong Kong dollars.

(b) No adjustment shall be made to the Final Contract Sum on account of any variation in the exchange rate between the Hong Kong dollar and any other currency.

(5) The Contract shall be governed by and construed in all respects according to the laws for the time being in force in Hong Kong.

Corresponding clauses in the ICE Conditions:

Clause 1 (*continued*)

(2) Words importing the singular also include the plural and vice-versa where the context requires.

(3) The headings and marginal notes in the Conditions of Contract shall not be deemed to be part thereof or be taken into consideration in the interpretation or construction thereof or of the Contract.

(4) All references herein to clauses are references to clauses numbered in the Conditions of Contract and not to those in any other document forming part of the Contract.

(5) The word "cost" when used in the Conditions of Contract means all expenditure properly incurred or to be incurred whether on or off the Site including overhead finance and other charges properly allocatable thereto but does not include any allowance for profit.

(6) Communications which under the Contract are required to be "in writing" may be hand-written typewritten or printed and sent by hand post telex cable facsimile or other means resulting in a permanent record.

Corresponding clauses in the FIDIC Conditions:

Clause 1.2

In the Contract, except where the context requires otherwise:

(a) words indicating one gender include all genders;

(b) words indicating the singular also include the plural and words indicating the plural also include the singular;

(c) provisions including the word "agree," "agreed" or "agreement" require the agreement to be recorded in writing, and

(d) "written" or "in writing" means hand-written, type-written, printed or electronically made, and resulting in a permanent record.

The marginal words and other headings shall not be taken into consideration in the interpretation of these Conditions.

Clause 1.4

The Contract shall be governed by the law of the country (or other jurisdiction) stated in the Appendix to Tender.

If there are versions of any part of the Contract which are written in more than one language, the version which is in the ruling language stated in the Appendix to Tender shall prevail.

條款 1

定義及解釋

（續）

(2) 如條文內容有所需要，指單數的詞也包括其眾數，反過來亦一樣。

(3) 「合同一般條件」、「合同特別條件」（如有的話）及「規格」的索引、旁註或標題，不可被視為這些條件及規格的部分，在解釋這些條件及規格時，也不可作為考慮。

(4) (a) 除另有規定外，所有付款必須以港元繳交。

(b) 不可因港元與任何其他貨幣匯率上的變動而調整「最終合同金額」。

(5) 本「合同」在各方面都必須為當時在香港實施的法律所規限，並必須按照此等法律作解釋。

Comparison with the ICE Conditions and the FIDIC Conditions

Term in HKSAR Conditions	Points to note in definition and interpretation		
	HKSAR Conditions	ICE Conditions	FIDIC Conditions
Marginal notes or Headings	Meaning similar, they shall not be taken into consideration.		
Governing laws	Laws of Hong Kong	Laws of England, Scotland and Northern Ireland	Laws of the country stated in the Appendix to the Tender.
Currency of payment	Hong Kong dollars	Local currency	Specified in the Particular Conditions with one currency as required by the financing institutions and also a local currency.

Clause 2

The HKSAR Conditions:

Duties and Powers of the Engineer and the Engineer's Representative

(1) (a) The Engineer shall carry out the duties and may exercise the powers specified in or necessarily to be implied from the Contract.

    (b) Before carrying out any such duty or exercising any such power, the Engineer may be required under the terms of his appointment by the Employer to obtain confirmation that the Employer has no objection to the Engineer's proposed course of action and, in the event of an objection, to act in accordance with the Employer's direction. If the Engineer is subject to any such requirements, particulars thereof shall be set out in the Appendix to the Form of Tender.

    (c) The Contractor's rights under the Contract shall not be prejudiced in any way by any failure on the part of the Engineer to comply with the requirements particularized in the Appendix to the Form of Tender or any other requirements of his appointment by the Employer.

    (d) Except as expressly stated in the Contract, the Engineer shall have no power to amend the terms and conditions of the Contract nor to relieve the Contractor of any of his obligations under the Contract.

(2) The duties of the Engineer's Representative are to watch and inspect the Works, to test and examine any material to be used and workmanship employed by the Contractor in connection with the Works and to carry out such duties and exercise such powers vested in the Engineer as may be delegated to him by the Engineer in accordance with the provisions of sub-clause (3) of this Clause.

Corresponding clauses in the ICE Conditions:

Clause 2

(1) (a) The Engineer shall carry out the duties specified in or necessarily to be implied from the Contract.

(b) The Engineer may exercise the authority specified in or necessarily to be implied from the Contract. If the Engineer is required under the terms of his appointment by the Employer to obtain the specific approval of the Employer before exercising any such authority particulars of such requirements shall be those set out in the Appendix to the Form of Tender. Any requisite approval shall be deemed to have been given by the Employer for any such authority exercised by the Engineer.

(c) Except as expressly stated in the Contract the Engineer shall have no authority to amend the Contract nor to relieve the Contractor of any of his obligations under the Contract.

(d) The giving of any consent or approval by or on behalf of the Engineer shall not in any way relieve the Contractor of any of his obligations under the Contract or of his duty to ensure the correctness or accuracy of the matter or thing which is the subject of the consent or approval.

(3) (a) The Engineer's Representative shall be responsible to the Engineer who shall notify his appointment to the Contractor in writing.

(b) The Engineer's Representative shall watch and supervise the construction and completion of the Works. He shall have no authority.

    (i) to relieve the Contractor of any of his duties or obligations under the Contract nor except as expressly provided for in sub-clause (4) of this Clause

    (ii) to order any work involving delay or any extra payment by the Employer or

    (iii) to make any variation of or in the Works.

Corresponding clauses in the FIDIC Conditions:

Clause 3.1

The Employer shall appoint the Engineer who shall carry out the duties assigned to him in the Contract. The Engineer's staff shall include suitably qualified engineers and other professionals who are competent to carry out these duties.

The Engineer shall have no authority to amend the Contract.

The Engineer may exercise the authority attributable to the Engineer as specified in or necessarily to be implied from the Contract. If the Engineer is required to obtain the approval of the Employer before exercising a specified authority, the requirements shall be as stated in the Particular Conditions. The Employer undertakes not to impose further constraints on the Engineer's authority, except as agreed with the Contractor.

However, whenever the Engineer exercises a specified authority for which the Employer's approval is required, then (for the purposes of the Contract) the Employer shall be deemed to have given approval.

Except as otherwise stated in these Conditions:

(a) whenever carrying out duties or exercising authority, specified in or implied by the Contract, the Engineer shall be deemed to act for the Employer;

(b) the Engineer has no authority to relieve either Party of any duties, obligations or responsibilities under the Contract; and

(c) any approval, check, certificate, consent, examination, inspection, instruction, notice, proposal, request, test, or similar act by the Engineer (including absence of disapproval) shall not relieve the Contractor from any responsibility for error, omissions, discrepancies.

條款 2

香港特別行政區土木工程合同一般條件：

「工程師」及「工程師代表」的職責與權力

(1) (a) 「工程師」必須履行在「合同」內指定或隱含的職責，並可使用在「合同」內指定或隱含的權力。

　　(b) 「工程師」在履行任何職責或使用任何權力前，「僱主」可在委任「工程師」的委任條款內規定「工程師」必須預先得到「僱主」確認不反對「工程師」所建議採取的行動。如有反對，「工程師」必須按照「僱主」的指示行事。如「工程師」要受上述規定所限制，此等規定的詳情必須在「投標表格」的「附件」內列出。

　　(c) 「承建商」在「合同」上的權利，不會因「工程師」不遵照在「投標表格」的「附件」內所列出的規定或不遵照「工程師」受「僱主」委任的任何其他規定而受到任何損害。

　　(d) 除「合同」明文闡述外，「工程師」無權修訂「合同」的條款及條件，亦無權減免「承建商」在「合同」上的任何責任。

(2) 「工程師代表」的職責是監督及視察「本工程」，測試及檢查「承建商」在「本工程」使用的任何物料及運用的技術水平，以及執行「工程師」獲授予並根據本條第(3)款的條文轉授予他的職責及使用授權予他的權力。

Commentary

Usually the Engineer under the Contract has a dual role. Firstly, he acts as the agent for the Employer in the management of the Contract and in supervising the Works. In this respect, he is acting in the interest of the Employer. Secondly, he must act impartially and not be biased against the Contractor in administering the Contract and in giving his direction and decision regarding material and workmanship, valuation and payment, etc. However, under 1(b) of this Clause, the Employer may impose constraints on the power of the Engineer. The current policy is that the Engineer must obtain the agreement of the Employer before ordering variation or committing the Government to expenditure in excess of HK$300,000, apart from claims. The Engineer is also required to report to the Employer all claims and variation, as well as the principles underlying his assessment of each claim, for the Employer to provide his view before the Engineer reaches a decision. If the Contract is a Mega Project Contract, e.g., with cost exceeding HK$1 billion, further constraints may be imposed on the power of the Engineer on approval of the programme, granting extension of time, handing over of the Site, etc., in accordance with the then Works Bureau Technical Circular No. 26/2002. As regards supervising the quality of material and workmanship, the Engineer may delegate his duties and power to the Engineer's Representative who then shall act on behalf of the Engineer in such matters.

Analysis and application

Clause 2(1)(a) "The Engineer shall carry out the duties … specified in or necessarily to be implied from the Contract."

In administering the Contract, the Engineer has to treat the Contractor with reasonable care and if he does not, he may breach his implied duty, to either the Employer or the Contractor or both. Even if the Contractor does not have a direct contract with the Engineer, they have a special relationship and the tort of negligence[1] may be committed by the Engineer, especially with respect to negligent mis-statement. The Engineer has also an implied duty to act fairly as between the Employer and the Contractor.[2]

Clause 2(1)(b) "… the Engineer may be required under the terms of his appointment by the Employer to obtain confirmation that the Employer has no objection to the Engineer's proposed course of action and, in the event of objection, to act in accordance with the Employer's direction …"

This has been discussed in the above commentary and technically the need to obtain confirmation and act upon the Employer's direction is called "referable decisions." Such requirements are written in the consultancy agreement between the Employer and the Engineer.

Clause 2(1)(c) "The Contractor's rights under the Contract shall not be prejudiced by any failure on the part of the Engineer to comply with the requirements … by the Employer."

The requirements may lengthen the time for the Contractor to get a decision from the Engineer who may also act against his own discretion according to the Employer's direction. However, if the Engineer does not refer the matter to the Employer in accordance with sub-clause 1(b), the Contractor is not required to make sure that the Engineer has done so and his right will not be affected.

Clause 2(1)(d) "… the Engineer shall have no power to amend the terms and conditions of the Contract nor to relieve the Contractor of any of his obligations under the Contract."

Even without such express provision, the Engineer has no ostensible or implied authority to vary the Contract or the Works by reason of his position as the Engineer.[3]

Clause 2(2) "The duties of the Engineer's Representative are to watch and inspect the Works, to test and examine any material to be used and workmanship employed by the Contractor …"

This sub-clause is self-explanatory. The Engineer's Representative is appointed to continually supervise the construction of the Works and he usually has only negative powers, to condemn bad work or material.[4]

 The Traffic Control and Surveillance System (TCSS) contractor, who is employed by the Government, lays the ducts. What is the status of the TCSS contractor under the contract? Can you find out from the definition in the Conditions of Contract?

 For the construction of the back-span of a cable-stayed bridge, specially designed falsework and temporary concrete columns are constructed by the sub-contractor. Define the responsibility of the contractor if the system collapses.

The HKSAR Conditions:

Duties and Powers of the Engineer and the Engineer's Representative

(3) The Engineer may from time to time delegate to the Engineer's Representative any of the duties and powers vested in him. Any such delegation shall be in writing signed by the Engineer and shall specify the duties and powers thereby delegated. No such delegation shall have effect until a copy thereof has been delivered to the Contractor. Any written instruction or written approval given by the Engineer's Representative to the Contractor within the terms of such delegation, but not otherwise, shall bind the Contractor and the Employer as though it had been given by the Engineer.

Provided that:

(a) failure of the Engineer's Representative to disapprove any work or material shall not prejudice the power of the Engineer thereafter to disapprove such work or material;

(b) if the Contractor or the Employer shall be dissatisfied by reason of any decision of the Engineer's Representative they may refer the matter to the Engineer who shall confirm, reverse or vary such decision.

(4) No act or omission by the Engineer or the Engineer's Representative in the performance of any of his duties or the exercise of any of his powers under the Contract shall in any way operate to relieve the Contractor of any of the duties, responsibilities, obligations or liabilities imposed upon him by any of the provisions of the Contract.

Corresponding clauses in the ICE Conditions:

Clause 2 *(continued)*

(4) The Engineer may from time to time delegate to the Engineer's Representative or any other person responsible to the Engineer any of the duties and authorities vested in the Engineer and he may at any time revoke such delegation. Any such delegation

(a) shall be in writing and shall not take effect until such time as a copy thereof has been delivered to the Contractor or his agent appointed under Clause 15(2);

(b) shall continue in force until such time as the Engineer shall notify the contractor in writing that the same has been revoked;

(c) shall not be given in respect of any decision to be taken or certificate to be issued under Clauses 12(6) 44 46(3) 48 60(4) 61 65 or 66.

(5) (a) The Engineer or the Engineer's Representative may appoint any number of persons to assist the Engineer's Representative in the carrying out of his

duties under sub-clauses (3)(b) or (4) of this Clause. He shall notify to the Contractor the names duties and scope of authority of such persons.

(b) Such assistants shall have no authority to issue any instructions to the Contractor save insofar as such instructions may be necessary to enable them to carry out their duties and to secure the acceptance of materials and workmanship as being in accordance with the Contract. Any instructions given by an assistant for these purposes shall where appropriate be in writing and be deemed to have been given by the Engineer's Representative.

(c) If the Contractor is dissatisfied by reason of any instruction of any assistant of the Engineer's Representative appointed under sub-clause (5)(a) of this Clause he shall be entitled to refer the matter to the Engineer's Representative who shall thereupon confirm reverse or vary such instruction.

Corresponding clauses in the FIDIC Conditions:

Clause 3.2

The Engineer may from time to time assign duties and delegate authority to assistants, and may also revoke such assignment or delegation. These assistants may include a resident engineer, and/or independent inspectors appointed to inspect and/or test items of Plant and/or Materials. The assignment, delegation or revocation shall be in writing

and shall not take effect until copies have been received by both Parties. However, unless otherwise agreed by both Parties, the Engineer shall not delegate the authority to determine any matter in accordance with Sub-Clause 3.5 [*Determinations*].

Assistants shall be suitably qualified persons, who are competent to carry out these duties and exercise this authority, and who are fluent in the language for communications defined in Sub-Clause 1.4 [*Law and Language*].

Each assistant, to whom duties have been assigned or authority has been delegated, shall only be authorized to issue instructions to the Contractor to the extent defined by the delegation. Any approval, check, certificate, consent, examination, inspection, instruction, notice, proposal, request, test, or similar act by an assistant, in accordance with the delegation, shall have the same effect as though the act had been an act of the Engineer. However:

(a) any failure to disapprove any work, Plant or Materials shall not constitute approval, and shall therefore not prejudice the right of the Engineer to reject the work, Plant or Materials;

(b) if the Contractor questions any determination or instruction of an assistant, the Contractor may refer the matter to the Engineer, who shall promptly confirm, reverse or vary the determination or instruction.

條款 2 （續）

香港特別行政區土木工程合同一般條件：

「工程師」及「工程師代表」的職責與權力

(3) 「工程師」可不時將他獲授予的任何職責及權力授權給「工程師代表」。任何此類授權必須以書面作出，由「工程師」簽署，並必須述明藉此授權的職責及權力。此類授權必須在授權書的副本交給承建商後才能生效。「工程師代表」所發給承建商的任何書面指令或書面批核，在此類授權條件範圍內，但不在其他方面，對「承建商」及「僱主」皆有約束力，而該指令或批核等同由「工程師」所發給的。但是：

(a) 「工程師代表」即使未能拒絕批核任何工程或物料，亦不會損害「工程師」日後拒絕批核此等工程或物料的權力；

(b) 如「承建商」或「僱主」不滿「工程師代表」的任何決定，他們可以將有關事情提交「工程師」，由「工程師」確認、推翻或更改此項決定。

(4) 「工程師」或「工程師代表」在履行他在「合同」內的任何職責或行使「合同」內的任何權力時，他的任何作為或不作為，皆不會在任何方面減免「合同」內任何條文加諸於「承建商」的任何職責、任務、責任或法律責任。

Analysis and application

Clause 2(3) "The Engineer may from time to time delegate to the Engineer's Representative any of the duties and powers vested in him …"

From this sentence, it appears that the HKSAR Conditions do not put any limitation on the duties to be delegated by the Engineer. In practice, the valuation of variation, the certification of claims, the granting of extension of time and the giving of decision on a matter under dispute are not delegated, as in the ICE Conditions.

Clause 2(3) "… Any written instruction or written approval given by the Engineer's Representative to the Contractor within the terms of such delegation, but not otherwise, shall bind the Contractor and the Employer as though it had been given by the Engineer …"

It should be noted that the Engineer has no power to amend the terms and conditions of the Contract, nor to relieve the Contractor of his obligations under the Contract under sub-clause 1(d) and the Engineer's Representative's instruction or approval on such aspects shall not have any effect.

Clause 2(3)(a) "failure of the Engineer's Representative to disapprove any work or materials shall not prejudice the power of the Engineer thereafter to disapprove such work or material;"

Please also see sub-clause (4), which says any act or omission by the Engineer or the Engineer's Representative in the performance of any of his duties shall not relieve the Contractor's obligations. Neither the Engineer nor the Engineer's Representative has the power to alter the Specification, unless such work is required under a variation[5] under Clause 60.

Clause 2(3)(b) "if the Contractor or the Employer shall be dissatisfied by reason of any decision of the Engineer's Representative, they may refer the matter to the Engineer who shall confirm, reverse or vary such decision."

The sub-clause is self-explanatory. It quite often happens on the Site that the Contractor may not be satisfied with the decision of the Engineer's Representative, e.g., payment matters and he then writes to the Engineer for a review first. If he is still not satisfied with the Engineer's review, he can seek decision under Clause 86.

Clause 2(4) "No act or omission by the Engineer or the Engineer's Representative in the performance of any of his duties or the exercise of any of his powers under the Contract shall in any way operate to relieve the Contractor of any of the duties ... imposed upon him by any provision of the Contract."

As explained in the commentary to sub-clause (3)(a), the Engineer has no power to alter the Contract requirements unless necessitated by a variation under Clause 60. If the Engineer approves material or workmanship not satisfying the Contract requirements, the Contractor can protect himself from future disapproval by requesting a variation order from the Engineer.

Comparison with the ICE Conditions

This ICE Clause 2 is similar to the HKSAR Clause except that it expressly states that the Engineer's Representative cannot order any work involving delay or any extra payment by the Employer or make any variation of the Works, and in particular he cannot delegate to the Engineer's Representative the power under Clauses 12(6), 44, 46(3), 48, 60(4), 61, 65 or 66 regarding granting extension of time, accelerating completion, certifying completion, issue of final and maintenance certificates and determination of the Contractor's appointment respectively.

Comparison with the FIDIC Conditions

One major difference between the FIDIC Conditions and the HKSAR and ICE Conditions is that Clause 3.2 states that the Engineer shall be deemed to act for the Employer. This means that the Engineer no longer acts in an impartial role between the Employer and the Contractor. If there is any requirement to obtain the Employer's approval before the Engineer exercises his authority, it should be stated in the Particular Conditions. The Engineer has no authority to relieve either Party from his duties, obligations and responsibility. Also, the Engineer's approval, certification, instruction or similar acts shall not relieve any of the Contractor's responsibility. The Engineer may delegate his duties to several "assistants," including a resident engineer and independent inspectors who must be suitably qualified. Failure of the assistants to disapprove workmanship or materials shall not prejudice the right of the Engineer to reject them and if the Contractor is not satisfied with the decision of the assistants, he can appeal to the Engineer.

Notes

1. *Day v Ost* [1973] 2NZLR, see also Note 2.
2. *Sutcliffe v Thakrah* [1974] AC 727. For a full discussion, see Max Abrahamson, *Engineering Law and the ICE Contracts*, 4th Edition, Chapters 13–14.
3. See the commentary in Ian Duncan Wallace, *A Commentary on the FIDIC International Standard Form of Engineering and Building Contract*, p. 28.
4. Max Abrahamson, *Engineering Law and the ICE Contracts*, 4th Edition, p. 34.
5. Max Abrahamson, *Engineering Law and the ICE Contracts*, 4th Edition, pp. 73 and 127.

The ground originally intended to be hydroseeded are over-compacted by plant. Should the Engineer give an order to loosen the ground? If the Engineer had given such an order, should the Contractor be paid for it?

Clause 5

Documents Mutually Explanatory

(1) Save to the extent that any Special Condition of Contract provides to the contrary the provisions of these General Conditions of Contract shall prevail over those of any other document forming part of the Contract.

(2) Subject to the foregoing the several documents forming the Contract are to be taken as mutually explanatory of one another but in case of ambiguities or discrepancies the same shall be explained by the Engineer who shall issue to the Contractor instructions clarifying such ambiguities or discrepancies. Where the Contractor makes a request in writing to the Engineer for instructions under this sub-clause the Engineer shall respond within 14 days of receipt of such request.

Provided that:

(a) work shown on the Drawings or described in the Specification but not measured in the Bills of Quantities shall be dealt with in accordance with Clause 59;

(b) if in the opinion of the Engineer compliance with such instructions shall involve the Contractor in any expense which by reason of any ambiguity or discrepancy the Contractor did not and had no reason to anticipate, the Engineer shall value such expense in accordance with Clause 61, and shall certify in accordance with Clause 79;

(c) if in the opinion of the Engineer compliance with such instructions shall involve the Contractor any saving then the Engineer shall value such saving and deduct the same from the Contract Sum accordingly.

Corresponding clauses in the ICE Conditions:

Clause 5

The several documents forming the Contract are to be taken as mutually explanatory of one another and in case of ambiguities or discrepancies the same shall be explained and adjusted by the Engineer who shall thereupon issue to the Contractor appropriate instructions in writing which shall be regarded as instructions issued in accordance with Clause 13.

Clause 13

(3) If in pursuance of Clause 5 or sub-clause (1) of this Clause the Engineer shall issue instructions which involve the Contractor in delay or disrupt his arrangements or methods of construction so as to cause him to incur cost beyond that reasonably to have been foreseen by an experienced contractor at the time of tender then the Engineer shall take such delay into account in determining any extension of time to which the Contractor is entitled under Clause 44 and the Contractor shall subject to Clause 53 be paid in accordance with Clause 60 the amount of such cost as may be reasonable except to the extent that such delay and extra cost result from the Contractor's default. Profit shall be added thereto in respect of any additional permanent or temporary work. If such instructions require any variation to any part of the Works the same shall be deemed to have been given pursuant to Clause 51.

Corresponding clauses in the FIDIC Conditions:

Clause 1.5

The documents forming the Contract are to be taken as mutually explanatory of one another. For the purposes of interpretation, the priority of the documents shall be in accordance with the following sequence:

(a) the Contract Agreement (if any),

(b) the Letter of Acceptance,

(c) the Letter of Tender,

(d) the Particular Conditions,

(e) these General Conditions,

(f) the Specification,

(g) the Drawings, and

(h) the Schedules and any other documents forming part of the Contract.

If an ambiguity or discrepancy is found in the documents, the Engineer shall issue any necessary clarification or instruction.

條款 5

文件的互相解釋

(1) 除「合同特別條件」內的任何條文有相反意思外，本「合同一般條件」的條文凌駕於任何作為「合同」部分的其他文件。

(2) 在符合上文規定下，作為本「合同」的各文件必須視作可互相解釋，但如出現含糊或差異時，則必須由「工程師」解釋，而「工程師」須向「承建商」發出指令，澄清含糊或差異之處。如「承建商」以書面要求「工程師」根據本款發出指令，「工程師」必須在收到這要求後14天內回覆。

但是：

(a) 在「圖則」上示明或「規格」內敘述的工程，如沒有在「工程量清單」內計量，則必須根據第59條處理；

(b) 如「工程師」認為依從上述指令會引致「承建商」牽涉任何開支，而此項開支是因任何含糊或差異造成，而「承建商」事前沒有亦沒有理由預計得到，則「工程師」必須根據第61條為此項開支釐訂價值，並且必須根據第79條予以核實；

(c) 如「工程師」認為依從以上指令將會引致「承建商」有任何節約，「工程師」必須為該項節約釐訂價值，並就此從「合同款項」中扣除該項節約。

Commentary

A typical Government construction contract includes basically the Conditions of Contract, Specification, Bill of Quantities, together with the Method of Measurement and Drawings. The *General Conditions of Contract*, *General Specification* and *Standard Method of Measurement* published by the Government are all standardized documents but amendments and additions to these documents are made from time to time to make improvements and to reflect Government policies through technical circulars or circular memoranda. Particular Specification, Particular Preamble to Standard Method of Measurement, Bill of Quantities and Drawings are prepared for individual contracts usually by the Consulting Engineers. For major contracts, these could come to tens of thousands of pages but are prepared within a few months. There are bound to be inconsistencies, say, within the drawings, or conflicts between specifications and drawings, or conditions of contract and specifications. This Clause attempts to deal with such conflicts and provides remedies to the Contractor.

Analysis and application

Clause 5(1) "Save to the extent that any Special Conditions of Contract provides to the contrary …"

The Special Conditions must give the extent to which it takes precedence over the General Conditions. In other words, to avoid argument, it must state which clauses or sub-clauses of the General Conditions are omitted or replaced by the Special Conditions. The Special Conditions must be vetted by lawyers.

Clause 5(1) "… the provisions of these General Conditions of Contract shall prevail over those of any other documents forming part of the Contract."

This accords a higher status to the General Conditions of Contract, which overrides all other documents in case of any inconsistencies and conflicts, except the Special Conditions of Contract. It is appreciated, however, that the General Conditions of Contract is a standard document, which describes the obligations, duties, responsibilities and liabilities of the contracting parties and the duties and powers of the Engineer. It is a framework and basis upon which all other documents are prepared and it rarely touches upon details regarding payment from which most contractual disputes arise. Direct conflicts between certain sentences or phrases of the General Conditions of Contract and other documents, such as the General Specification, do not usually occur, especially when the other documents are prepared by an experienced drafter.

Clause 5(2) "… the several documents forming the Contract are to be taken as mutually explanatory of one another but in case of ambiguities or discrepancies the same shall be explained by the Engineer …"

It suggests that the other documents such as the Specification, Bill of Quantities and Drawings are all accorded with equal status. By virtue of definition in Clause 1, the Specification, Bill of Quantities and Drawings includes all post-contract documents incorporated by variation orders or supplemental agreements. It is often necessary to find out the true intentions of the parties from different parts of the contract documents, e.g., for multiple ducts surrounded by concrete, do we measure the length of only one duct or the length of all individual ducts and add all the lengths together for the purpose of payment? To find out the true intention, we have to read the Standard Method of Measurement, the Particular Preamble to it, the Drawings, the Bill of Quantities and sometimes the Specification. To find out the true meaning, we can follow the Court's general rule of interpretation such as the Golden Rule[1] (using plain ordinary meaning, e.g., it is arguable that the surface area of a glass panel means one side, two sides or six sides of it), customs and technical usage (based on expert's opinion), and *Ejusdem Generis* (any general words following particular words of a similar class refers to that class, e.g., soil, rocks, boulders and other materials where other materials should usually refer to natural materials like stones, pebbles but not steel). If two or more possible interpretations are possible, the Engineer has to pay attention to the Contra Proferentum rule, which means the interpretation shall be against the party responsible for drafting the clause. Usually legal advice is required for interpretation having major significance as regards to cost or time. Sometimes, the Court would imply a term into the contract for business efficacy, but a term will only be implied if it is not inconsistent with the express terms.[2]

Clause 5(2) "… the Engineer who shall issue to the Contractor instructions clarifying such ambiguities or discrepancies … within 14 days of receipt of such request …"

Upon request by the Contractor or if the Engineer considers necessary without a request, the Engineer shall issue instruction to clarify ambiguities or discrepancies. Many contractors make systematic requests for further information and clarifications in the hope of making out cases for seeking reimbursement or extension of time and the Engineer may find himself inundated by such requests. The Engineer should bear in mind that the Contractor shall usually be responsible for supplying construction materials complying with the Specification and choosing his own method of construction. He may point out that many of such requests are in fact not necessary. If such practice persists, he should bring it up to the Contractor's higher management.

Clause 5(2)(a) "work shown on the Drawings or described in the Specification but not measured in the Bills of Quantities shall be dealt with in accordance with Clause 59;"

This refers to missing items, which are to be dealt with under Clause 59(3) and the value of work involved shall be ascertained in accordance with Clause 61(1)(a).

Clause 5(2)(b) "if in the opinion of the Engineer compliance with such instruction shall involve the Contractor in any expense … the Engineer shall value such expense in accordance with Clause 61, …"

The word "expense" is not defined in the Contract but it is similar to the measure of damages, which arise naturally in the usual course of things. It includes Site overhead and office overhead but it is unclear whether profit should be included. However, since the valuation is based on Bills of Quantities rates in accordance with Clause 61, profit has been included.

Comparison with the ICE Conditions

The ICE Conditions accords equal status to all contract documents and therefore in case of conflict, more weight could be attached to those documents specifically prepared for the Contract as a whole, e.g., the Bills of Quantities or Particular Specification rather than the printed *Conditions of Contract* or *Standard Method of Measurement* (which are of general application and may be less likely to reflect the intention of the parties when entering into the Contract).

Comparison with the FIDIC Conditions

The FIDIC Conditions provides a hierarchy between the various documents forming the contract. It is to be noted that Part II "Conditions of Particular Application" provides alternative forms of wording, where other order of precedence might be preferred, or alternatively where it is decided that no order of precedence should be included, in which case it is similar to the ICE Conditions.

Notes

1. *Liverpool City Council v Irwin* [1976] All ER39.
2. *Trollope Colls Ltd. v North West Metropolitan Regional Hospital Board* [1973] All ER 260.

 The Specification requires that the minimum clear spacing between these bolts shall be 75mm. The Drawing shows that the bolts are 75mm apart centre to centre. What should the Contractor do?

The HKSAR Conditions:

Provision of Drawings and Specification

(1) Two copies of the Contract and two additional copies of the Drawings shall be furnished to the Contractor free of charge.

(2) The Engineer shall within 14 days of receiving a request in writing from the Contractor provide the Contractor with any further copies of the Drawings requested by the Contractor upon payment of the standard rate laid down from time to time by the Employer.

(3) The Engineer shall issue to the Contractor from time to time during the progress of the Works such other Drawings and Specification as in the opinion of the Engineer shall be necessary for the purpose of the execution of the Works and the Contractor shall be bound by the same.

(4) The Contractor shall give adequate notice in writing to the Engineer of other Drawings or Specification that may be required for the execution of the Works.

The HKSAR Conditions:

Information Not to be Divulged

(1) The Contractor shall not use or divulge, except for the purpose of the Contract, any information provided by the Employer or the Engineer in the Contract or in any subsequent correspondence or documentation.

Corresponding clauses in the ICE Conditions:

Clause 6

(1) Upon award of the Contract the following shall be furnished to the Contractor free of charge:
(a) four copies of the Conditions of Contract Specification and (unpriced) Bill of Quantities and
(b) the number and type of copies as entered in the Appendix to the Form of Tender of all Drawings listed in the Specification.

Clause 6

(3) Copyright of all Drawings Specifications and the Bill of Quantities (except the pricing thereof) supplied by the Employer or the Engineer shall not pass to the Contractor but the Contractor may obtain or make at his own expense any further copies required by him for the purposes of the Contract. Similarly copyright in all documents supplied by the Contractor under Clause 7(6) shall remain with the Contractor but the Employer and the Engineer shall have full power to reproduce and use the same for the purpose of completing operating maintaining and adjusting the Works.

Clause 7

(1) The Engineer shall from time to time during the progress of the Works supply to the Contractor such modified or further Drawings Specifications and instructions as shall in the Engineer's opinion be necessary for the purpose of the proper and adequate construction and completion of the Works and the Contractor shall carry out and be bound by the same.

Corresponding clauses in the FIDIC Conditions:

Clause 1.8

The Specification and Drawings shall be in the custody and care of the Employer. Unless otherwise stated in the Contract, two copies of the Contract and of each subsequent Drawing shall be supplied to the Contractor, who may make or request further copies at the cost of the Contractor.

Each of the Contractor's Documents shall be in the custody and care of the Contractor, unless and until taken over by the Employer. Unless otherwise stated in the Contract, the Contractor shall supply to the Engineer six copies of each of the Contractor's Documents.

The Contractor shall keep, on the Site, a copy of the Contract, publications named in the Specification, the Contractor's Documents (if any), the Drawings and Variations and other communications given under the Contract. The Employer's Personnel shall have the right of access to all these documents at all reasonable times.

If a Party becomes aware of an error or defect of a technical nature in a document which was prepared for use in executing the Works, the Party shall promptly give notice to the other Party of such error or defect.

Clause 1.11

As between the Parties, the Employer shall retain the copyright and other intellectual property rights in the Specification, the Drawings and other documents made by (or on behalf of) the Employer. The Contractor may, at his cost, copy, use, and obtain communication of these documents for the purposes of the Contract. They shall not, without the Employer's consent, be copied, used or communicated to a third party by the Contractor, except as necessary for the purposes of the Contract.

條款 6

香港特別行政區土木工程合同一般條件：

「圖則」與「規格」之供應

(1) 「承建商」須得到免費供應「合同」副本兩份及額外「圖則」副本兩份。

(2) 「工程師」必須在收到「承建商」書面要求額外「圖則」副本後14天內供應給「承建商」，並按照「僱主」不時訂下的標準收費率收取費用。

(3) 在「本工程」施行期間，「工程師」須不時發放給「承建商」他認為實施「本工程」所需要的其他「圖則」及「規格」，而「承建商」必須受此等「圖則」及「規格」約束。

(4) 「承建商」如為實施「本工程」需要其他「圖則」或「規格」，必須給予「工程師」充分的書面通知。

條款 8

香港特別行政區土木工程合同一般條件：

不可洩露資料

(1) 除了為「合同」的作用外，「承建商」不可使用或洩露「僱主」或「工程師」在「合同」或任何日後的通信或文件中供應的資料。

Commentary

It is not uncommon that when a civil engineering contract is let, the tenderers are provided with generalized Drawings sufficient to enable the tenderers to price the Bill of Quantities. During the course of Contract, the Contractor is issued with "working drawings," which give the details of the works to be executed. Alternatively, the Contractor may be required under the Contract to submit detailed working drawings for the Engineer's approval. For example, the general arrangement drawings of concrete structures are given at the tender stage and the reinforced concrete drawings are given during the course of construction. For steelworks, the general arrangement drawings are given during the tender while the shop drawings are either issued by the Engineer or to be submitted by the Contractor for approval. If the Contractor or sub-contractor is required to produce shop drawings or working drawings, it is essential to define the responsibility for design, or Specifications as between the Employer and the Contractor.[1] This Clause, together with Clause 7, govern the manner and the timing in which such Drawings are issued or submitted. The issue of Drawings under this Clause must be distinguished from the issue of Drawings under Clause 60 since the latter is intended for variation to the Works.

Analysis and application

Clause 6(3) "The Engineer shall issue to the Contractor from time to time during the progress of the Works such other Drawings and Specification as in the opinion of the Engineer shall be necessary for the purpose of the execution of the Works …"

This Clause imposes a duty on the Engineer to issue detailed Drawings to the Contractor to enable him to carry out the works, normally in accordance with his latest programme or 3-month rolling programme. Nowadays, most Government departments require their engineers or consultants to complete virtually all the reinforcement Drawings and other detailed Drawings before tendering except when tendering under a very tight programme. In the latter case, the Engineer should preferably complete all detailed Drawings during the early part of the Contract to avoid the Contractor submitting a claim under Clause 50 for extension of time and under Clause 63 for disturbances to the progress of Works. Though sub-clause (4) requires the Contractor to give adequate notice to the Engineer for the further Drawings, it is submitted that such "adequate notice" may be construed as only a couple of weeks and would not be long enough for the Engineer to carry out the design and checking, and prepare the Drawings. This Clause also applies to Specification. The Contractor may request further details on Specification, especially concerning proprietary products or equivalent. Sometimes it is very difficult to decide who should be responsible for the Specification and this should preferably be explicitly stated in the Contract.

Clause 6(3) "… and the Contractor shall be bound by the same."

The Contractor is under an obligation to carry out the Works in accordance with the Drawings issued by the Engineer during the construction period. It is submitted that if the details given by such Drawings are very difficult to construct, e.g., complicated shear links in the slabs, which cannot be foreseeable from the Bill of Quantities, Specification, etc., the Contractor may submit a claim.

Clause 6(4) "The Contractor shall give adequate notice in writing to the Engineer of other Drawings or Specifications that may be required for the execution of the Works."

Apparently, this Clause protects the Engineer from breaching his duty to supply detailed Drawings to suit the Contractor's programme by requiring the Contractor to give adequate notice. However, as explained in the preceding paragraph, adequate notice may be construed as only a couple of weeks. As also explained in the commentary to Clause 5, many contractors make systematic requests for further information/clarification, making the job of the Engineer extremely difficult. While every request should be responded to, the Engineer must be able to focus on the ones that have potentially serious contractual implications in order to avoid the Contractor making a huge claim on extra management charges and disruption costs.[2]

Clause 8(1) "The Contractor shall not use or divulge, except for the purpose of the Contract, any information provided by the Employer or the Engineer in the Contract or in any subsequent correspondence or documentations."

This Clause, together with Clause 36 regarding photographs, prohibits the Contractor from using the Contract information for practice promotion or any other commercial purposes without the consent of the Employer.

Comparison with the ICE Conditions

The ICE Clause refers to "modified or further Drawings Specification and instructions … necessary for the purpose of the proper and adequate construction and completion of the Works …" and may suggest that the Clause intends to cover variations as well. But by referring to ICE Conditions sub-clause 7(4) on p. 26, any compensation under this Clause will only cover any delay in issuing further Drawings, Specifications and instructions but not extension of time or costs arising from modified Drawings or variations. Regarding construction information, the ICE Clause expressly refers to copyright which is the current approach in protecting any misuse of such information by the Contractor.

Comparison with the FIDIC Conditions

The FIDIC Clause does not impose an express duty upon the Engineer to issue further Drawings but largely relies on "the Contractor shall give notice to the Engineer whenever the Works are likely to be delayed or disrupted if any necessary Drawing or instruction is not issued within a particular time …" (Clause 1.9 on p. 26) This may be appropriate since the requirements of international contracts may vary. The notice shall include "details of the necessary drawing or instruction, details why and by when it should be issued, and details of the nature and amount of delay or disruption …" This may to some extent prevent the Contractor from making systematic requests. Regarding Contract information, the FIDIC Clause refers to both copyright and other intellectual property rights which is the current approach in protecting against any misuse of information by the Contractor.

Notes

1. *Fairweather & Co. v London Borough of Wandsworth* (1987) 39BLR106. See also Roger Knowles, *150 Contractual Problems and Their Solutions*, Section 1.9.
2. See also the commentary on Clause 63, p. 159.

 5 The noise enclosure is designed by the Contractor based on the general layout and setting out Drawings of the bridge structure provided by the Engineer. However, because of segmental construction, the cross falls of the structure are not as accurate as envisaged and the structural frames do not match. Who should be responsible?

The HKSAR Conditions:

Provision of Drawings and Specification

(5) One copy of the Drawings furnished to the Contractor as aforesaid shall be kept by the Contractor on the Site and the same shall at all reasonable times be available for inspection and use by the Engineer and the Engineer's Representative and by any other person authorized by the Engineer in writing.

(6) At the Completion of the Works the Contractor, if required by the Engineer, shall return to the Engineer all Drawings and other Contract documents provided under the Contract, other than the Contractor's signed copy of such Drawings or documents.

Corresponding clauses in the ICE Conditions:

Clause 7

(3) The Contractor shall give adequate notice in writing to the Engineer of any further Drawing or Specification that the Contractor may require for the construction and completion of the Works or otherwise under the Contract.

(4) (a) If by reason of any failure or inability of the Engineer to issue at a time reasonable in all the circumstances Drawings Specifications or instructions requested by the Contractor and considered necessary by the Engineer in accordance with sub-clause (1) of this Clause the Contractor suffers delay or incurs additional cost then the Engineer shall take such delay into account in determining any extension of time to which the Contractor is entitled under Clause 44 and the Contractor shall subject to Clause 53 be paid in accordance with Clause 60 the amount of such cost as may be reasonable.

(b) If the failure of the Engineer to issue any Drawing Specification or instruction is caused in whole or in part by the failure of the Contractor after due notice in writing to submit drawings specifications or other documents which he is required to submit under the Contract the Engineer shall take into account such failure by the Contractor in taking any action under sub-clause (4)(a) of this Clause.

Corresponding clauses in the FIDIC Conditions:

Clause 1.9

The Contractor shall give notice to the Engineer whenever the Works are likely to be delayed or disrupted if any necessary drawing or instruction is not issued to the Contractor within a particular time, which shall be reasonable. The notice shall include details of the necessary drawing or instruction, details of why and by when it should be issued, and details of the nature and amount of the delay or disruption likely to be suffered if it is late.

If the Contractor suffers delay and/or incurs cost as a result of failure of the Engineer to issue the notified drawing or instruction within a time which is reasonable and is specified in the notice with supporting details, the Contractor shall give further notice to the Engineer and shall be entitled subject to Sub-Clause 20.1 [*Contractor's Claims*] to:

(a) an extension of time for any such delay, if completion is or will be delayed, under Sub-Clause 8.4 [*Extension of Time for Completion*], and

(b) payment of any such cost plus reasonable profit, which shall be included in the Contract Price.

After receiving this further notice, the Engineer shall proceed in accordance with Sub-Clause 3.5 [*Determinations*] to agree or determine these matters.

However, if and to the extent that the Engineer's failure was caused by any error or delay by the Contractor, including an error in, or delay in the submission of, any of the Contractor's documents, the Contractor shall not be entitled to such extension of time, cost or profit.

香港特別行政區土木工程合同一般條件：

「圖則」與「規格」之供應

（續）

(5) 「承建商」必須在工地內存有上文所提及供應給他的「圖則」副本各一份，給「工程師」及「工程師代表」，以及「工程師」書面授權的任何人士在任何合理時間查察及使用。

(6) 在「本工程」竣工後，如「工程師」要求，「承建商」必須發還給「工程師」所有按照「合同」得到的「圖則」及其他「合同」文件，但不包括「承建商」簽署的復印本。

Analysis and application

Clause 6(5) "… The same shall at all reasonable times be available for inspection and use by the Engineer …"
This Clause, if used at all, shall be for audit purposes to make sure that the Contractor keeps and uses the latest copy of the Drawings for the construction of the Works.

Comparison with the ICE Conditions

The ICE Clause specifically specifies the remedy in extension of time and cost compensation if the Contractor suffers delay as a result of the Engineer's failure to issue at a time reasonable in all circumstances Drawings, Specifications or instructions requested by the Contractor. For the HKSAR Conditions, the Contractor has to rely on Clause 63(a). The ICE Clause also expressly provides for the Engineer to take into account the Contractor's failure to give adequate notice in granting extension of time and certifying payment.

Comparison with the FIDIC Conditions

Only if the Contractor suffers delay and incurs cost as a result of the Engineer's failure to issue the notified Drawings or instruction within a reasonable time and the Contractor has given a prior notice with supporting details, the Contractor is then in a position to give further notice to the Engineer for claiming extension of time and cost. The FIDIC Clause also allows the Engineer to take into account the Contractor's fault in granting extension of time, certifying cost and profit. The FIDIC Clause allows for reasonable profit to be made by the Contractor.

Clause 7

The HKSAR Conditions:

Drawings Provided by the Contractor for the Works

(1) When the Contractor is required to provide Drawings or other documents in connection with the Works, unless the Contract provides to the contrary, all such Drawings and documents shall be submitted in duplicate to the Engineer at a reasonable time before the work shown or described thereon is to be carried out so as to permit the Engineer sufficient time to examine the Contractor's proposals properly. The Engineer shall give or refuse his approval in writing to such proposals within a reasonable time.

(2) If the Engineer has reasonable cause for being dissatisfied with the proposals set out in the Contractor's Drawings or documents the Engineer shall require the Contractor to make such amendments thereto as the Engineer may consider reasonably necessary. The Contractor shall make and be bound by such amendments at no additional expense to the Employer.

(3) The Contractor shall provide the Engineer with the type and number of copies of such Drawings and documents as may be specified in the Contract within 14 days of the Engineer's approval.

(4) Should it be found at any time after approval has been given by the Engineer that the details do not comply with the terms and conditions of the Contract or that the details do not agree with the Drawings or documents previously submitted and approved by the Engineer, the Contractor shall make such alterations or additions as in the opinion of the Engineer are necessary to remedy such non-compliance or non-agreement at the Contractor's own expense.

(5) No examination by the Engineer of the Drawings or documents submitted by the Contractor under the provisions of this Clause nor any approval given by the Engineer of the same, with or without amendment, shall absolve the Contractor from any liability for the same.

Clause 8

(continued)

The HKSAR Conditions:

Information Not to be Divulged

(2) The Employer and the Engineer may use any information provided by the Contractor in accordance with the Contract but shall not divulge such information except for the purpose of the Contract or for the purpose of carrying out any repair, amendment, extension or other work connected with the Works.

Corresponding clauses in the ICE Conditions:

Clause 7

(2) Where sub-clause (6) of this Clause applies the Engineer may require the Contractor to supply such further documents as shall in the Engineer's opinion be necessary for the purpose of the proper and adequate construction completion and maintenance of the Works and when accepted by the Engineer the Contractor shall be bound by the same.

(6) Where the Contract expressly provides that part of the Permanent Works shall be designed by the Contractor he shall submit to the Engineer for acceptance

(a) such drawings specifications calculations and other information as shall be necessary to satisfy the Engineer that the Contractor's design generally complies with the requirements of the Contract and

(b) operation and maintenance manuals together with completed drawings of that part of the Permanent Works in sufficient detail to enable the Employer to operate maintain dismantle reassemble and adjust the Permanent Works incorporating that design. No certificate under Clause 48 covering any part of the Permanent Works designed by the Contractor shall be issued until manuals and drawings in such detail have been submitted to and accepted by the Engineer.

(7) Acceptance by the Engineer in accordance with sub-clause (6) of this Clause shall not relieve the

Contractor of any of his responsibilities under the Contract. The Engineer shall be responsible for the integration and co-ordination of the Contractor's design with the rest of the Works.

Corresponding clauses in the FIDIC Conditions:

Clause 1.8

Each of the Contractor's Documents shall be in the custody and care of the Contractor, unless and until taken over by the Employer. Unless otherwise stated in the Contract, the Contractor shall supply six copies of each of the Contractor's Documents.

If the party becomes aware of an error or defect of a technical nature in a document which was prepared for use in executing the Works, the Party shall promptly give notice to the other Party such error or defect.

Clause 1.10

As between the Parties, the Contractor shall retain the copyright and other intellectual property rights in the Contractor's Documents and other design documents made by (or on behalf of) the Contractor.

The Contractor shall be deemed (by signing the Contract) to give to the Employer a non-terminable transferable non-exclusive royalty-free licence to copy, use and communicate the Contractor's Documents, including making and using modifications of them. This licence shall:

(a) apply throughout the actual or intended working life (whichever is longer) of the relevant parts of the Works,

(b) entitle any person in proper possession of the relevant part of the Works to copy, use and communicate the Contractor's Documents for the purposes of completing, operating, maintaining, altering, adjusting, repairing and demolishing the Works, and

(c) in the case of Contractor's Documents which are in the form of computer programs and other software, permit their use on any computer on the Site and other places as envisaged by the Contract, including replacements of any computers supplied by the Contractor.

The Contractor's Documents and other design documents made by (or on behalf of) the Contractor shall not, without the Contractor's consent, be used, copied or communicated to a third party by (or on behalf of) the Employer for purposes other than those permitted under this Sub-Clause.

條款 7

香港特別行政區土木工程合同一般條件：

「承建商」為「本工程」供應的「圖則」

(1) 當「承建商」有需要提供與「本工程」有關的「圖則」或其他文件時，除「合同」另有規定外，「承建商」必須在此等「圖則」或文件所示明或敘述的工程開始前的一個合理時間內，提交全部一式兩份給「工程師」，讓「工程師」有足夠時間妥善地審閱「承建商」的建議。「工程師」必須在合理時間內以書面批核或拒絕批核這些建議。

(2) 工程師如有合理的理由對「承建商」的「圖則」或文件上所列出的建議不滿意，「工程師」必須要求「承建商」作出「工程師」認為需要的合理修訂。「承建商」必須作出這些修訂及受其修訂約束，「僱主」亦無須為這些修訂負擔額外費用。

(3) 「承建商」必須在得到「工程師」批核後14天內，向「工程師」供應「合同」所指明的種類及數量的「圖則」副本及文件副本。

(4) 在「工程師」批核後的任何時間內，如發現細節不遵照「合同」內的條款及條件，或細節和以前提交並經「工程師」批核的「圖則」或文件不一致，「承建商」必須自費作出「工程師」認為需要的修改或增補，用以補救上述不遵照或不一致的情況。

(5) 對於「承建商」按照本條文所提交的「圖則」或文件，雖經「工程師」審閱或批核（無論是否有修訂），但仍不免除「承建商」對這些「圖則」或文件所負的法律責任。

條款 8

不可洩露資料

（續）

(2) 「僱主」及「工程師」可以使用「承建商」按照「合同」供應的任何資料，但除了為「合同」的作用或因進行與「本工程」有關的任何修補、修訂、伸延或其他與「本工程」有關的工程外，不可將資料洩露。

Commentary

This Clause governs the procedures, which the Contractor has to provide Drawings to the Engineer for approval before carrying out the Works. Such Drawings may include, for example, detailed shop drawings for steel work, or pipework for water points, or detailed drawings for noise barrier sub-frames, etc. If the Contractor has to design part of the Works specified in the Contract, he shall submit both general arrangement and detailed Drawings. Generally, the timing for submission by the Contractor and approval by the Engineer must be reasonable. The Engineer can require amendments so far he has reasonable cause for being dissatisfied. Even after the approval has been given, any further amendments necessary to remedy any non-compliance with the Contract shall be at the Contractor's expense.

Analysis and application

Clause 7(4) "Should it be found at any time after approval has been given by the Engineer that details do not comply with the terms and conditions of the Contract … the Contractor shall make such alterations … to remedy such non-compliance … at the Contractor's own expense."

This echoes the principle in Clause 2(4) that "No act or omission by the Engineer … in the performance of any of his duties … shall in any way operate to relieve the Contractor of any of the duties … imposed upon him by any of the provisions of the Contract."

Clause 7(5) "No examination by the Engineer of the Drawings or documents submitted by the Contractor … shall absolve the Contractor from any liability for the same."

This sub-clause 7(5) merely serves to reiterate the above principle in Clause 2(4).

Clause 8(2) "The Employer and the Engineer may use any information provided by the Contractor … but shall not divulge such information except for the purpose of the Contract or for the purpose of carrying out … extension or other work connected with the Works."

This sub-clause allows the Employer to use the Contractor's Drawings for the maintenance, operations and future extension or modification of the Works.

Comparison with the ICE Conditions

The ICE Clause 7 governs the details of submission for the Permanent Works which expressly are required to be designed by the Contractor. It does not cover temporary works whose requirements are described in sub-clauses 14(6) and 14(7). Further requirements for the temporary works may be specified in other documents such as special conditions or particular specification. For the HKSAR Government contracts, the detailed requirements for the temporary works are specified in a Special Conditions of Contract. For the permanent works designed by the Contractor, the ICE Clause requires that it is "necessary to satisfy the Engineer that the Contractor's design generally complies with the requirements of the Contract" and not necessarily suitable or adequate, and this ensures that the responsibility remains at all times with the Contractor. However, the Engineer is responsible for the integrating and co-ordinating of the Contractor's design with the rest of the Works.

Comparison with the FIDIC Conditions

The FIDIC Clause 1.8 does not say very much about the detailed procedures regarding Drawings submitted by the Contractor. However, the last paragraph of Clause 1.8 requires each Party to give the other Party prompt notice when the former becomes aware of an error or defect of a technical nature in a document. This aims at minimizing the effect of any error in a document. The *Notes on the Preparation of Tender Documents* recommends that the requirements for the Contractor's documents should be included in the particular specification. Similarly, in the HKSAR Government contracts, detailed requirements for submission of contractor's documents for both temporary and permanent works are usually specified in the special conditions or in the particular specification.

The FIDIC Clause 1.10 explains in detail that the Contractor retains the copyright of his documents but allows the Employer to copy, use and modify the Contractor's documents in the form of a licence.

 The Contractor uses specially designed falsework for the cross heads of the back-span of the cable-stayed bridge. In the event that many years after, the back-span has to be extended, can the Employer make use of those specially designed falsework Drawings?

Clause 13

Inspection of the Site

(1) The Contractor shall be deemed to have examined and inspected the Site and its surroundings and to have satisfied himself, before submitting his Tender, as regards existing roads or other means of communication with and access to the Site, the nature of the ground and sub-soil, the form and nature of the Site, the risk of injury or damage to property, the nature of materials (whether natural or otherwise) to be excavated, the nature of the work and materials necessary for the execution of the Works, the accommodation he may require, and generally to have obtained his own information on all matters affecting his Tender and the execution of the Works.

(2) No claim by the Contractor for additional payment shall be allowed on the ground of any misunderstanding in respect of the matters referred to in sub-clause (1) of this Clause or otherwise or on the ground of any allegation or fact that incorrect or insufficient information was given to him by any person whether in the employ of the Employer or not or of the failure of the Contractor to obtain correct and sufficient information, nor shall the Contractor be relieved from any risk or obligation imposed on or undertaken by him under the Contract on any such ground or on the ground that he did not or could not foresee any matter which may in fact affect or have affected the execution of the Works.

Corresponding clauses in the ICE Conditions:

Clause 11

(2) The Contractor shall be deemed to have inspected and examined the Site and its surroundings and information available in connection therewith and to have satisfied himself so far as is practicable and reasonable before submitting his tender as to
(a) the form and nature thereof including the ground and sub-soil and hydrological conditions and
(b) the extent and nature of work and materials necessary for constructing and completing the Works and
(c) the means of communication with and access to the Site and the accommodation he may require

and in general to have obtained for himself all necessary information as to risks contingencies and all other circumstances which may influence or affect his tender.

Corresponding clauses in the FIDIC Conditions:

Clause 4.10

The Employer shall have made available to the Contractor for his information, prior to the Base Date, all relevant data in the Employer's possession on sub-surface and hydrological conditions at the Site, including environmental aspects. The Employer shall similarly make available to the Contractor all such data which come into the Employer's possession after the Base Date. The Contractor shall be responsible for interpreting all such data.

To the extent which was practicable (taking account of cost and time), the Contractor shall be deemed to have obtained all necessary information as to risks, contingencies and other circumstances which may influence or affect the Tender or Works. To the same extent, the Contractor shall be deemed to have inspected and examined the Site, its surroundings, the above data and other available information, and to have been satisfied before submitting the Tender as to all relevant matters, including (without limitation):

(a) the form and nature of the Site, including sub-surface condition,
(b) the hydrological and climatic conditions,
(c) the extent and nature of the Work and Goods necessary for the execution and completion of the Works and the remedying of any defects,
(d) the Laws, procedures and labour practices of the Country, and
(e) the Contractor's requirements for access, accommodation, facilities, personnel, power, transport, water and other services.

條款 13

視察「工地」

(1) 「承建商」必須被視為在提交「標書」前經已檢查及視察「工地」及其周圍環境，並已確定有關「工地」現有的道路或其他與「工地」聯絡或進入「工地」的途徑、土地及地下土壤的性質、「工地」的形狀及性質、人身或財產受損的風險、須挖掘的物件（無論是否天然物質）的性質、為實施「本工程」所需的工程及物料的性質、以及「承建商」或需要的住所。而且「承建商」亦必須被視為他已大致地就一切影響其「投標」及「本工程」實施的事物取得他所需要的資料。

(2) 「承建商」不可根據他對本條第(1)款所提及的事物或其他方面有任何誤解、或聲稱有或事實上有任何受僱於「僱主」或不受僱於「僱主」的人向他提供不正確或不足夠的資料、或「承建商」本身未能取得正確及足夠資料而索償額外付款。「承建商」亦不可根據上述理由、或因他沒有或未能預測到任何可能事實上影響或已影響「本工程」實施的事物為理由而減免「合同」加諸於他或他根據「合同」而必須承擔的任何風險和責任。

Commentary

This Clause attempts to place all the risks associated with the Site conditions on the Contractor since even though it may not be practicable to carry out thorough examination and investigation during the limited tender period, the Contractor is deemed to have done so as stated in sub-clause (1). Further, sub-clause (2) attempts to exempt the Employer from any liability arising from any incorrect information provided by him or those acting on his behalf, barring the Contractor from any financial remedy. However, this Clause has a number of problems. In determining the liability of the Parties, the Court may take into account the relatively short tender period available to the Contractor for examining and inspecting the Site or carrying out his own investigation[1] and the kind of representation made by the Employer or his agent (whether the misrepresentation is fraudulent, reckless or innocent,[2] or whether the representation by the Employer is sufficient to become a term of the contract or form a collateral contract[3]). It should also be noted that the exclusion clause is normally narrowly construed by the Court[4] and it would apply the Contra Proferentum rule sparingly against the drafter.

Analysis and application

Clause 13(1) "The Contractor shall be deemed to have examined and inspected the site and its surroundings and to have satisfied himself … the nature of the ground and sub-soil … the nature of materials … to be excavated …"

In the HKSAR Government contracts, the risk of unexpected site or ground conditions falls upon the Contractor and the above deeming provision attempts to achieve this policy. The general principle of English Law is that the Contractor has an obligation to deal with unexpected difficulties at his own expense, in the absence of express provisions to the contrary. As described in the comparison with the ICE Conditions or the FIDIC Conditions later, the HKSAR Conditions do not have a Clause 12 allowing financial compensation for adverse physical conditions and artificial obstructions. However, the Contractor seeking redress may sometimes rely on Clause 15 "physical impossibility" to avoid his obligation to complete or force the Engineer to issue variations under Clause 60 to overcome the impossibility. The Contractor also relies on Clause 50 to seek extension of time, and especially on sub-clause (1)(b)(xi) — "any special circumstance of any kind whatsoever" — which has an extremely wide wording and is likely to cover unforeseen ground condition or any uncharted utilities (the right to extension of time for uncharted utilities is now covered by a Special Condition). For this Clause 13 itself, it is submitted that while the Contractor is required to inspect the Site, his duty to inspect may not extend to opening up the ground to investigate the sub-soil or locating underground utilities. The Contractor must at least satisfy himself by inspecting excavations in the neighbourhood, making enquires from local contractors and examination of geological and other records. It has been argued that while the Contractor has to satisfy himself as regards "the nature of material to be excavated," it does not extend to piling work.

Clause 13(1) "… and generally to have obtained his own information on all matters affecting his Tender and the execution of the Works."

While the Contractor is required "generally to have obtained his own information," it is a misconception that the Employer shall withhold site investigation reports,[5] other geological information or utilities information in order to avoid claims from the Contractor. Such reports or utilities plans are prepared by specialist contractors or utilities undertakings based on objective data for the purpose of design and construction of the Works and would be invaluable to the Contract as a whole. Provided that the information is genuine and reasonable disclaimer clauses are attached, it is unlikely that the Contractor can succeed in a claim against the Employer solely because the actual ground condition deviates from such reports. For interpretative reports which are subjective, e.g., ground appraisal or recommended traffic diversion plans prepared during the planning stage, it is submitted that the Employer should avoid releasing to the Contractor during the tender stage even with disclaimer, since such information may form part of the Contract and any future changes to suit Site or ground condition could amount to a variation. However, such information may be released post-contract on a "without admission of liability" basis since the Contractor does not rely on it for his pricing during tender.

Clause 13(2) "No claim by the Contractor for additional payment shall be allowed on the ground of any misunderstanding in respect of the matters referred to in sub-clause (1) … or on the ground of any allegation or fact that incorrect or insufficient information was given to him …"

This sub-clause (2) is an exemption clause disclaiming liability for giving incorrect information. An exemption clause is given strict interpretation by the Court. It only excludes implied terms and the Contra Proferentum rule will apply. It, therefore, will not protect the Employer when other express terms are provided to the contrary intention, as described on p. 33. The exemption clause will not protect the Employer if there is fraud or misrepresentation.[6] Exemption clause may sometimes be subject to reasonableness test. As regards "insufficient information," however, since there is no misrepresentation, liability would not arise.[7]

Clause 13(2) "… given to him by any person whether in the employ of the Employer or not …"

The usual person to give information to the tenderers is the Consulting Engineer who will act as the Engineer for the Contract during the construction stage. The Employer may be bound if the Engineer has express or ostensible authority to given information, which amounts to a misrepresentation under the *Misrepresentation Ordinance* (Cap 284).

Clause 13(2) "… nor shall the Contractor be relieved from any risk or obligation imposed on or undertaken by him under the Contract on any such ground or on the ground that he did not or could not foresee any matter which may in fact affect or have affected the execution of the Works."

As described previously on p. 33, there is no clause in the HKSAR Conditions, which allows for adverse physical conditions and artificial obstructions and the above wording states this explicitly.

Comparison with the ICE Conditions

This ICE Clause is similar to the HKSAR Clause 13(1) as far as the various aspects of the Site and other things, which the Contractor shall be deemed to have inspected. It does not include the nature of material to be excavated but include hydrological condition. The former is not included possibly because the Contractor can rely on the ICE Clause 12 to claim compensation for adverse physical conditions or artificial obstructions, which could not have been foreseen by an experienced Contractor. Under this ICE Clause 11, the Contractor only needs to have satisfied himself "as far as is practicable" and it is submitted that this requirement is more reasonable and compatible with the limited period during tendering, e.g., the Contractor may not be expected to carry out extensive ground investigation during the tender period. The inclusion of geological condition is because, unlike the HKSAR Clause 50, the ICE Clause 44 only allows extension of time for exceptional adverse weather condition and the tenderer must assess how his future work would be affected by the normal local weather condition. The ICE Clause 11 does not have an exemption clause for incorrect or insufficient information since it has the ICE Clause 12 for unforeseen conditions, as explained above.

Comparison with the FIDIC Conditions

The second paragraph of this FIDIC Clause 4.10 is very similar to the ICE Clause (and also the HKSAR Clause) but includes further aspects such as Laws, labour practices, power, transport, etc., because of its international nature. The first paragraph of this FIDIC Clause puts an obligation on the Employer to release all relevant data in the Employer's possession on sub-surface and hydrological condition, including environmental aspects. In Hong Kong, most major projects must have made submission in accordance with the *Environmental Impact Assessment Ordinance* (EIAO) and the submission would be incorporated into the tender documents.

Notes

1. *Morrison-Krudson Int. v Commonwealth of Australia* [1972].
2. *Pearson & Son Ltd. v Dublin Corporation* [1907] AC 351.
3. *Bascal Construction (Midlands) Ltd. v Northampton Development Corporation* [1975].
4. *Andrews Bros. (Bournemouth) Ltd. v Singer & Co. Ltd.* [1934] 1KB17.
5. *Opron Construction Co. v Alberta* (1994) C.L.R. (2d) 97 (Alta, QB), p. 298. See also Roger Knowles, *150 Contractual Problems and Their Solutions*, Section 2.6.
6. *Curtis v Chemical Cleaning and Dyeing Co.* [1951] 1All ER631 and also the *Misrepresentation Ordinance*.
7. *Keates v Lord Cadogan* [1851] 10CB591.

On the left side is a works site for an elevated road to be constructed and this is a picture taken during the weekend. On a weekday, the road is fully occupied by container vehicles tailing back from the container storage area and blocking access to the works site. The progress of construction is seriously affected. Can the Contractor have any remedy?

Clause 14

Sufficiency of Tender

The Contractor shall be deemed to have satisfied himself before submitting his Tender as to the correctness and sufficiency of his Tender for the Works and of the rates stated in the priced Bills of Quantities which rates shall, except in so far as it is otherwise provided in the Contract, cover all his risks, liabilities and obligations set out or implied in the Contract and all matters and things necessary for the proper execution of the Works.

Corresponding clauses in the ICE Conditions:

Clause 11

(3) The Contractor shall be deemed to have
(a) based his tender on his own inspection and examination as aforesaid and on all information whether obtainable by him or made available by the Employer and

(b) satisfied himself before submitting his tender as to the correctness and sufficiency of the rates and prices stated by him in the Bill of Quantities which shall (unless otherwise provided in the Contract) cover all his obligations under the Contract.

Corresponding clauses in the FIDIC Conditions:

Clause 4.11

The Contractor shall be deemed to:
(a) have satisfied himself as to the correctness and sufficiency of the Accepted Contract Amount, and
(b) have based the Accepted Contract Amount on the data, interpretations, necessary information, inspections, examinations and satisfaction as to all relevant matters referred to in Sub-Clause 4.10 [*Site Data*].

Unless otherwise stated in the Contract, the Accepted Contract Amount covers all the Contractor's obligations under the Contract (including those under Provisional Sums, if any) and all things necessary for the proper execution and completion of the Works and the remedying of any defects.

條款 14

充分滿意的「標書」

在提交「標書」前，「承建商」必須被視為已滿意他就「本工程」提交的「標書」及已落價的「工程量清單」所載錄的價率的正確及充分。除「合同」另有規定外，這些價率必須已包括在「合同」內列出或隱含須由「承建商」承擔的一切風險、法律責任及其他責任，以及包括為妥善地實施「本工程」所需要的一切事物及東西。

Commentary

This Clause complements Clause 13 under which the Contractor is to be deemed to have examined the Site and generally have obtained his own information on all matters affecting his Tender. The Clause requires the Contractor to have satisfied himself to the correctness and sufficiency of his Tender and the rates he puts in shall cover all his risks, liabilities and obligations expressly stated or implied in the Contract. However, some of the liabilities and risks may be shared by the Employer. For example, Clause 59(3) requires the Engineer to value the works in connection with error in the description of the Bill of Quantities or item omitted. Claims and disputes may arise because under Clause 59(1), the Bill of Quantities shall be deemed to have been prepared and measurements made according to the Method of Measurement incorporated into the Contract. Together with the need to issue variation orders necessary for the completion of the Works, this Clause 14 is not so forceful in practice as its wording appears to be.

Analysis and application

Clause 14 "… the rates stated in the priced Bill of Quantities which rates shall, except in so far as it is otherwise provided in the Contract, cover … all matters and things …"

The Contract provides for additional payment arising from errors in the description of the Bill of Quantities, missing items or substantial change in quantities under Clause 59, variation order under Clause 60, disturbance to the progress of work under Clause 63, other reimbursement such as non-possession of the Site under Clause 48, and excepted risks under Clause 20, etc.

Clause 14 "… cover all his risks, liabilities and obligations set out or implied in the Contract and all matters and things necessary for the proper execution of the Works."

The "obligations … implied in the Contract" can cover obligations implied by law, customs and trade practices. There has always been argument whether any work or things included in the Specification and the Drawings but not included in the Bill of Quantities by way of item coverage (i.e., the description of work or things to be carried out under the item) shall be regarded as missing items or otherwise. As long as the current Standard Method of Measurement is used, the argument will persist. The Development Bureau is currently reviewing the Standard Method of Measurement with a view to eliminating or minimizing such argument.

Comparison with the ICE Conditions

The ICE sub-clause (3)(b) is much simplified but conveys essentially the same meaning as the HKSAR Clause.

Comparison with the FIDIC Conditions

This FIDIC Clause refers to the correctness and sufficiency of the "Accepted Contract Amount" instead of rates in the priced Bill of Quantities. It should be noted that under the re-measurement contract in the FIDIC Conditions, the Accepted Contract Amount is not a Lump Sum but is built up from the rates and prices in the Bill of Quantities.

The HKSAR Conditions:

Works to be to the Satisfaction of the Engineer

Save in so far as it is legally or physically impossible the Contractor shall execute the Works in strict accordance with the Contract to the satisfaction of the Engineer and shall comply with and adhere strictly to the Engineer's instructions on any matter related to the Contract whether mentioned in the Contract or not.

Corresponding clauses in the ICE Conditions:

Clause 13

(1) Save insofar as it is legally or physically impossible the Contractor shall construct and complete the Works in strict accordance with the Contract to the satisfaction of the Engineer and shall comply with and adhere strictly to the Engineer's instructions on any matter connected therewith (whether mentioned in the Contract or not). The Contractor shall take instructions only from the Engineer or subject to Clause 2(4) from his duly appointed delegate.

(2) The whole of the materials Contractor's Equipment and labour to be provided by the Contractor under Clause 8 and the mode manner and speed of construction of the Works are to be a kind and conducted in a manner acceptable to the Engineer.

(3) If in pursuance of Clause 5 or sub-clause (1) of this Clause the Engineer shall issue instructions which involve the Contractor in delay or disrupt his arrangements or methods of construction so as to cause him to incur cost beyond that reasonably to have been foreseen by an experienced contractor at the time of tender then the Engineer shall take such delay into account in determining an extension of time to which the Contractor is entitled under Clause 44 and the Contractor shall subject to Clause 53 be paid in accordance with Clause 60 the amount of such cost as may be reasonable except to the extent that such delay and extra cost result from the Contractor's default. Profit shall be added thereto in respect of any additional Permanent or Temporary Works. If such instructions require any variation to any part of the Works the same shall be deemed to have been given pursuant to Clause 51.

Corresponding clauses in the FIDIC Conditions:

Clause 4.1 (*part*)

The Contractor shall design (to the extent specified in the Contract), execute and complete the Works in accordance with the Contract and with the Engineer's instructions, and shall remedy any defects in the Works.

Clause 19.7

Notwithstanding any other provision of this Clause, if any event or circumstance outside the control of the Parties (including, but not limited to, Force Majeure) arises which makes it impossible or unlawful for either or both Parties to fulfil its or their contractual obligations or which, under the law governing the Contract, entitles the Parties to be released from further performance of the Contract, then upon notice by either Party to the other Party of such event or circumstance:

(a) the Parties shall be discharged from further performance, without prejudice to the rights of either Party in respect of any previous breach of the Contract, and

(b) the sum payable by the Employer to the Contractor shall be the same as would have been payable under Sub-Clause 19.6 [*Optional Termination, Payment and Release*] if the Contract had been terminated under Sub-Clause 19.6.

條款 15

「本工程」須達至「工程師」滿意的程度

除在法律上或實際上不可能外，「承建商」必須嚴格按照「合同」實施「本工程」，以使到「工程師」滿意，並且必須嚴格遵守「工程師」對任何與「合同」有關的事宜的指令，無論這些事宜在「合同」中是否有提及。

Commentary

This Clause 15 imposes an obligation on the Contractor to complete the Works in strict accordance with the Contract to the satisfaction of the Engineer and comply with his instruction on any matters related to the Contract. This Clause appears to give very wide power to the Engineer in controlling the execution of the Works. Under the Contract, the Engineer can issue instruction to clarify documents and Drawings, order variations and if he is dissatisfied with any material, workmanship, method of construction, safety aspects or the progress of Works, he can request the Contractor to take all necessary measures to his satisfaction to remedy the situation. There are various provisions under the Contract under which the Contractor may claim extension of time and monetary compensation, e.g., Clause 5 for clarifying documents and Clause 60 for variation, but the Contractor's duty to complete the Works shall not be relieved even he faces great difficulties during construction, unless he can plead legal or physical impossibility. The issue concerning impossibility raises a lot of contentions, locally and abroad, and it is not uncommon for the Contractor to plead impossibility when he experiences difficulties with ground conditions, utilities or even high prices in carrying out the Works. In some cases, the Engineer may have to modify his design and issue variations to overcome the alleged impossibility.

Analysis and application

Clause 15 "Save in so far it is legally or physically impossible …"

Legal impossibility means the Works, whether in whole or in part, are prohibited by law or infringe private rights which are capable of being protected by injunction, e.g., works contravening the EIAO, infringing the intellectual property rights or intruding into private lands. For physical impossibility, it is much more difficult to define and this leads to a lot of contentions with far-reaching consequences. Physical impossibility could mean that the works are extremely difficult to construct, to the extent that even if a lot of additional works, extra plant and resources are employed, the Contractor is still unable to bring the works to completion. Physical impossibility could also mean some supervening event, such as landslide, or fire destroying the Site[1] and the Works, or the lack of construction material on a global basis making further construction impossible or impracticable,[2] and this may be dealt with under the common law of frustration[3] or Clause 85 which provides for frustrations. Whatever the cause of physical impossibility, it is submitted that with the current engineering technology, very few things cannot be achieved with ample time, effort and money, but the question is whether such results are within the original contemplation of the parties in a business context.[4] In a pipe-laying contract in which the specified tolerances were extremely stringent, the contractor plead physical impossibility. The Court held that absolute impossibility was not the meaning to be applied but the situation should be looked at from a "practical commercial point of view" and "in an ordinary competitive commercial basis."[5] This is in fact the principle being followed in the current litigations and arbitrations. For major HKSAR contracts tendered with a marking scheme, the Contractor's proposal submitted during the tender is usually bound into the Contract. A Special Condition requires the Contractor to rectify his own deficiencies in the design if the design is subsequently found to be impossible to construct. However, if the impossibility is not due to the design deficiencies but arising from supervening events, such as worldwide shortage of materials, the Contractor may not be responsible.

Clause 15 "… shall execute the Works in strict accordance with the Contract and to the satisfaction of the Engineer and shall comply with and adhere strictly to the Engineer's instructions on any matter related to the Contract whether mentioned in the Contract or not."

The Engineer can give instructions to clarify ambiguities and rectify errors in documents and order variations necessary for the completion of the Works but may not give instructions positively contrary to the Contract[6] or change the essential terms of the Contract, as explained further in the commentary to Clause 60. It should be noted that acts of the Engineer outside the provisions of the Contract would not bind the Employer under Clause 2(4) unless he has express or ostensible authority from the Employer. To some extent, the wording "to the satisfaction of the Engineer" gives the Engineer a say or an influential role on the method of construction. If alternative methods of construction are available under the Contract but with just one pricing mechanism, the Engineer may approve the method, which appears to the Engineer to be safer or with less disruption to the public, and the Contractor may not have any remedies against the Employer. However, the Engineer should be careful not to positively instruct the Contractor with regard to the method of construction in general.

Comparison with the ICE Conditions

This ICE Clause 13(1) is basically the same as the HKSAR Clause 15. It further elaborates that the instruction can be given by the Engineer's Representative subject to the ICE Clause 2(4). Regarding the Engineer's instruction, sub-clause (3) allows for compensation of time and cost if the instruction involves the Contractor in delaying or disrupting his working arrangements or method of construction.

Comparison with the FIDIC the Conditions

The FIDIC Clause 4.1 is similar to the HKSAR Clause 15, but the FIDIC Clause includes the obligations in design to the extent specified in the Contract, which is often applicable to many contracts. In the HKSAR contracts, a Standard Special Conditions of Contract provides for any design obligation of the Contractor. The wording "to the satisfaction of the Engineer" has been omitted since it may be of little legal significance eventually with the adjudication and arbitration provisions under the FIDIC Clauses 20.4 and 20.6. The FIDIC Clause 19.7 provides relief to both the Contractor and the Employer and specifies explicitly that the circumstances leading to impossibility must be outside the control of the Parties. It also specifies the effect of impossibility on the obligations of the Parties.

Notes

1. *Taylor v Caldwell* (1863) 3B&S.
2. *Jackson v Union Marine Insurance Co. Ltd.* (1874) LR10; *Davis Contractors Ltd. v Fareham UDC* [1956] AC 696.
3. *Taylor v Caldwell* (1863) 3B&S.
4. *Tito v Waddell* [1977] 3All ER129.
5. *Turrif Ltd. v Welsh National Water Board* [1994] CLYB122.
6. Max Abrahamson, *Engineering Law and the ICE Contracts*, 4th Edition, p. 73.

Owing to the traffic, utilities and environmental constraints, the Contractor responsible for building the flyover has overspent his budget by several times and he is going to spend much more in order to complete the project. What is his contractual position?

Clause 16

Programme to be Furnished

(1) Within 14 days of the acceptance of the Tender or within such other period of time as may be specified in the Contract the Contractor shall submit to the Engineer a programme showing the sequence, method and timing, including (in so far as such work is described in the Contract) due allowance for the carrying out of Specialist Works and work by utility undertakings, in which the Contractor proposes to carry out the Works and shall, whenever required by the Engineer, furnish for the Engineer's information, particulars in writing of the Contractor's arrangements for carrying out the Works and of the Constructional Plant and Temporary Works which the Contractor intends to supply, use or construct as the case may be.

(2) The submission to the Engineer of such programme, or revised programme in accordance with Clauses 50 or 51, or the furnishing of such particulars shall not relieve the Contractor of any duty or responsibility under the Contract.

Corresponding clauses in the ICE Conditions:

Clause 14

(1) (a) Within 21 days after the award of the Contract the Contractor shall submit to the Engineer for his acceptance a programme showing the order in which he proposes to carry out the Works having regard to the provisions of Clause 42(1).

(b) At the same time the Contractor shall also provide in writing for the information of the Engineer a general description of the arrangements and methods of construction which the Contractor proposes to adopt for the carrying out of the Works.

(c) Should the Engineer reject any programme under sub-clause (2)(b) of this Clause the Contractor shall within 21 days of such rejection submit a revised programme.

(2) The Engineer shall within 21 days after receipt of the Contractor's programme

(a) accept the programme in writing or

(b) reject the programme in writing with reasons or

(c) request the Contractor to supply further information to clarity or substantiate the programme or to satisfy the Engineer as to its reasonableness having regard to the Contractor's obligations under the Contract.

Provided that if none of the above actions is taken within the said period of 21 days the Engineer shall be deemed to have accepted the programme as submitted.

(3) The Contractor shall within 21 days after receiving from the Engineer any request under sub-clause (2) (c) of this Clause or within such further period as the Engineer may allow provide the further information requested failing which the relevant programme shall be deemed to be rejected.

Upon receipt of such further information the Engineer shall within a further 21 days accept or reject the programme in accordance with sub-clauses (2)(a) or (2)(b) of this Clause.

(4) Should it appear to the Engineer at any time that the actual progress of the work does not conform with the accepted programme referred to in sub-clause (1) of this Clause the Engineer shall be entitled to require the Contractor to produce a revised programme showing such modifications to the original programme as may be necessary to ensure completion of the Works or any Section within the time for completion as defined in Clause 43 or extended time granted pursuant to Clause 44. In such event the Contractor shall submit his revised programme within 21 days or within such further period as the Engineer may allow. Thereafter the provisions of sub-clauses (2) and (3) of this Clause shall apply.

Corresponding clauses in the FIDIC Conditions:

Clause 8.3

The Contractor shall submit a detailed time programme to the Engineer within 28 days after receiving the notice under Sub-Clause 8.1 [*Commencement of Works*]. The Contractor shall also submit a revised programme whenever the previous programme is inconsistent with actual progress or with the Contractor's obligations. Each programme shall include:

(a) the order in which the Contractor intends to carry out the Works, including the anticipated timing of each stage of design (if any), Contractor's Documents, procurement, manufacture of Plant, delivery to Site, construction, erection and testing,

(b) each of these stages for work by each nominated Subcontractor (as defined in Clause 5 [*Nominated Subcontractors*]),

(c) the sequence and timing of inspections and tests specified in the Contract, and

(d) a supporting report which includes:

(i) a general description of the methods which the Contractor intends to adopt, and of the major stages, in the execution of the Works, and

(ii) details showing the Contractor's reasonable estimate of the number of each of Contractor's Personnel and of each type of Contractor's Equipment, required on the Site for each major stage.

Unless the Engineer, within 21 days after receiving a programme, gives notice to the Contractor stating the extent to which it does not comply with the Contract, the Contractor shall proceed in accordance with the programme, subject to his other obligations under the Contract. The Employer's Personnel shall be entitled to rely upon the programme when planning their activities.

The Contractor shall promptly give notice to the Engineer of specific probable future events or circumstances which may adversely affect the work, increase the Contract Price or delay the execution of the Works. The Engineer may require the Contractor to submit an estimate of the anticipated effect of the future event or circumstances, and/or a proposal under Sub-Clause 13.3 [*Variation Procedure*].

If at any time, the Engineer gives notice to the contractor that a programme fails (to the Extent stated) to comply with the Contract or to be consistent with actual progress and the Contractor's stated intentions, the Contractor shall submit a revised programme to the Engineer in accordance with this sub-clause.

條款 16

香港特別行政區土木工程合同一般條件：

提交施工計劃

(1) 在「標書」獲接納後14天內，或在「合同」指定的其他時間內，「承建商」必須向「工程師」提交一份施工計劃，在此計劃中示明「承建商」建議施行本工程的次序、方法及時間，包括考慮到「專門工程」及公用事業機構施行的工程(如「合同」描述此等工程)的次序、方法及時間。每當「工程師」提出要求時，「承建商」必須以書面向「工程師」提交詳情，敘述「承建商」為施行「本工程」所作出的安排以及「承建商」預定供應、使用或建造（視乎情況而定）的「施工設備」和「臨時工程」給「工程師」參考。

(2) 向「工程師」提交上述施工計劃，或按照第50或51條提交上述施工計劃，或經修改後的施工計劃或提交上述詳情，皆不會減免「承建商」在「合同」內的任何職責或責任。

Commentary

This Clause 16 requires the Contractor to submit to the Engineer a programme showing the sequence, method and timing for carrying out the Works and if requested by the Engineer to provide further information regarding the Contractor's working arrangement, Constructional Plant and Temporary Works. In the HKSAR Government contracts, this Clause is usually modified by Special Conditions and supplemented by Specification which give detailed requirements, including the use of critical path analysis and computer programme with software compatible to that used by the Engineer's site staff. The programme has to comply with all the contract requirements including availability of Portions of the Site, time for completion of various Sections and other constraints imposed by the Contract. This Clause 16 programme is generally referred to as detailed Works programme and in conjunction with it, there is the initial programme to be agreed in principle before the submission of the detailed Works Programme, the 3-month rolling programme for impending activities, the summary level programme for reporting purposes and the sub-programmes for major activities. It should be noted that the Clause 16 programme is a Contract Programme showing the intention of the Contractor as to his order and arrangement in carrying out the Works and is a very useful tool for the Engineer to manage the Contract by timely supply of information and Drawings to the Contractor, arranging his supervisory resources, and monitoring the timely completion of the Works. However, the programme has now been increasingly exploited by the Contractor, not as a management tool, but as a "contractual" programme for claiming extension of time and associated prolongation cost in case any event causes slippage to the Contractor's planned activities. The Engineer has the duty to see that the programme submitted by the Contractor must be realistic and logical, and supported by his planned and actually available resources. Non-approval of programme is very common but the Engineer must endeavour to agree with the Contractor as far as possible the majority part of the programme and give conditional approval where appropriate. If the agreed programme subsequently becomes unrealistic owing to major slippage or substantial extension of time, the Engineer should request the Contractor to submit a revised programme. In assessing the extension of time, the Engineer should not limit himself to what is contained in the agreed programme but should consider all the relevant factors such as Contractor's manner of working, the adequacy of resources, concurrency and float times as well as the Contractor's duty to mitigate delay. For major contracts, the Contractor has to submit monthly progress reports on all activities and the Engineer should check whether such progress reports reflect the reality. This is because such reports are commonly used in updating the progress to carry out time slice analysis for determining extension of time.[1] For programme or method statements required to be submitted during the tender stage, care must be taken to state clearly whether the programme or method statement shall become part of the contract document or the specified method of construction.[2]

Analysis and application

Clause 16(1) "… the Contractor shall submit to the Engineer a programme showing the sequence, method and timing, including … due allowance for the carrying out of Specialist Works and work by utility undertakings, …"

Sequence and timing are shown in bar charts together with critical path activity linkages. A major roadwork project can be broken down into more than 2,000 activities and therefore sophisticated software such as Primavera is necessary. A young engineer may have to spend several days, together with the guidance of the programmer, to understand the logic of the programme activities but it is worthwhile to spend the time. Experienced engineer may get a good grasp of the critical activities by examining the float times and the works flow analysis diagrams, both are readily available from the programme data. The Engineer can request that the bar chart programme be elaborated and explained by narrative statements. Sometimes, negative float times may appear in the programme which means the Contractor cannot meet the due date for completion and the Contractor actually implies that he is entitled to extension of time which has not been granted by the Engineer. The term "Specialist Works" refers to work by the Contractor employed by the Employer, which should be distinguished from the Approved List of Specialist Contractors. The former is now seldom employed in the HKSAR Government civil engineering contracts. The Engineer should request the Contractor to have early co-ordinations with utilities undertakings in order that utility sub-programmes are incorporated into the main programme, instead of simply allowing a window of a couple of weeks for utility works, which always turn out to be grossly inadequate.

Clause 16(1) "… and shall, whenever required by the Engineer, furnish for the Engineer's information, particulars in writing of the Contractor's arrangements for carrying out the Works and of the Constructional Plant and Temporary Works …"

The particulars are the Contractor's method statements for major activities, the Constructional Plant and Temporary Works to be employed for the Works, e.g., details of piling plant for piling works or launching girders for pre-cast segments for bridge deck construction.

Clause 16(2) "The submission to the Engineer of such programme, … or the furnishing of such particulars shall not relieve the Contractor of any duty or responsibility under the Contract."

The programme has to satisfy all the requirements and subject to all the constraints of the Contract. The Contractor shall be responsible for his own failure to incorporate all the requirements of the Contract, e.g., if the Contractor fails to allow for the diversion of an important water main which subsequently delay his critical works, he has no claim against the Employer.

Comparison with the ICE Conditions

This ICE Clause requires the submission of a programme to the Engineer "for his acceptance." The Engineer is required to either accept the programme or reject the programme with reasons, or request the Contractor to supply further information to clarify or substantiate the programme or to satisfy the Engineer as to its reasonableness. It is submitted that the ICE Clause imposes a more positive obligation on the Engineer to deal with the submission by the Contractor. It also sets a time limit of 21 days for the parties to respond to each other. For some major contracts of the HKSAR Government, such positive obligations are imposed on the Engineer by way of Special Conditions. It is also submitted that conditional approval of the programme should be expressly provided for in the Special Conditions to prevent acceptance of the programme being held up by disagreement over a part of the Works.

Comparison with the FIDIC Conditions

This FIDIC Clause 8.3 requires additionally the submission of the timing of design activities, the preparation of Contractor's Documents and also activities of the Nominated Sub-contractor. This FIDIC Clause also requires information regarding the Contractor's Personnel, which includes supervisory staff and labour. In the HKSAR Government contracts, such information is usually requested through Special Conditions or Particular Specification. This FIDIC Clause also put positive obligations on the Engineer to respond to Contractor's submission on the programme by requiring him to state the extent to which the Contractor's programme does not comply with the Contract within 21 days. This provision is similar to the corresponding ICE sub-clauses 14(2) and (3) but the ICE Clause is more comprehensive in this respect.

Notes

1. See also the commentary on sub-clause 50(3) on p. 101.
2. *Yorkshire Water Authority v Sir Alfred McAlpine (1985) Ltd.* 32BLR119; *English Industrial Estate Corporation v Kier Construction Ltd. and Others* (1991) 56BLR98. See also Roger Knowles, *150 Contractual Problems and Their Solutions*, Section 6.4.

 The Contractor can always increase his resources or change his method of construction to catch up with his own delay, but is unwilling to do so if the Employer or the Engineer causes the delay. How does the programme assist the Engineer to administer the Contract?

Clause 17

Contractor's Superintendence

(1) The Contractor shall give or provide all necessary superintendence during the execution of the Works and as long thereafter as the Engineer may consider necessary for the proper fulfilment of the Contractor's obligations under the Contract.

(2) The Contractor shall ensure that he is at all times represented on the Site by a competent and authorized English-speaking agent who shall be deemed to be approved by the Engineer provided such agent is not expressly disapproved by the Engineer in writing within 14 days from the serving of a notice in writing upon the Engineer by the Contractor of the appointment of such agent. Such agent shall be constantly on the Site and shall give his whole time to the superintendence of the Works.

(3) The Engineer shall have the authority to withdraw his approval of the authorized agent at any time. If such approval shall be withdrawn the Contractor shall, after receiving notice in writing of such withdrawal, remove the agent from the Site forthwith and shall not thereafter employ him again on the Site in any capacity and shall replace him by another competent English-speaking agent approved by the Engineer.

(4) Such authorized agent shall receive on behalf of the Contractor directions and instructions from the Engineer and the Engineer's Representative.

Corresponding clauses in the ICE Conditions:

Clause 15

(1) The Contractor shall provide all necessary superintendence during the construction and completion of the Works and for as long thereafter as the Engineer may reasonably consider necessary.

Such superintendence shall be given by sufficient persons having adequate knowledge of the operation to be carried out (including the methods and techniques required the hazards likely to be encountered and methods of preventing accidents) for the satisfactory and safe construction of the Works.

(2) The Contractor or a competent and authorized agent or representative approved of in writing by the Engineer (which approval may at any time be withdrawn) is to be constantly on the Works and shall give his whole time to the superintendence of the same. Such authorized agent or representative shall be in full charge of the Works and shall receive on behalf of the Contractor directions and instructions from the Engineer or (subject to the limitations of Clause 2) the Engineer's Representative. The Contractor or such authorized agent or representative shall be responsible for the safety of all operations.

Corresponding clauses in the FIDIC Conditions

Clause 6.8

Throughout the execution of the Works, and as long thereafter as is necessary to fulfil the Contractor's obligations, the Contractor shall provide all necessary superintendence to plan, arrange, direct, manage, inspect and test the work.

Superintendence shall be given by a sufficient number of persons having adequate knowledge of the language for communications (defined in Sub-Clause 1.4 [*Law and Language*]) and of the operations to be carried out (including the methods and techniques required, the hazards likely to be encountered and methods of preventing accidents), for the satisfactory and safe execution of the Works.

條款 17	*香港特別行政區土木工程合同一般條件：* 「承建商」的監督

(1) 在「本工程」實施期間，及其後在「工程師」認為「承建商」須妥善地完成「合同」內的責任所需要的期間內，「承建商」必須給予或提供一切所需要的監督。

(2) 「承建商」必須確保在任何時間內皆有一名稱職、獲授權並能說英語的代表作為他的駐「工地」代理人。「承建商」以書面通知「工程師」委任此代理人後，如「工程師」沒有在此通知送達後14天內以書面明文地不予批核，此代理人可被視為已獲「工程師」批核。此代理人必須常駐「工地」，並須以全部時間監督「本工程」。

(3) 「工程師」擁有隨時可撤銷對獲授權代理人的批核的權力。如此項批核被撤銷，「承建商」必須在接到關於撤銷此項批核的書面通知後，立刻將代理人遷出工地，並以後不得再僱用此代理人以任何身份在「工地」內工作。「承建商」必須用另一名稱職的、能說英語並獲「工程師」批核的代理人取代被撤換的代理人。

(4) 獲授權的代理人必須代「承建商」接收「工程師」及「工程師代表」發出的指示和指令。

Commentary

This Clause requires the Contractor to provide all necessary superintendence during the execution of the Works to fulfil his obligations under the Contract and to employ a competent and English-speaking agent as his representative. The minimum qualification of the agent is usually found in the Particular Specification. Depending on the size of the project, a recognized university degree is the minimum requirement. For a major project, a membership of the Hong Kong Institution of Engineers or Institutions of Civil Engineer is required. In Hong Kong, though the Site Agent is the authorized representative, a project manager who actually looks after the whole of the Works is present on Site and the Engineer's Representative deals with both of them. Most of the Contractor's correspondence are usually signed by the project manager.

Analysis and application

Clause 17 This Clause has a straightforward interpretation. In order to satisfy the Engineer regarding the provision of "all necessary superintendence … for the proper fulfilment of the Contractor's obligations under the Contract," the Contractor has to provide an organization chart and the detailed experience of each person.

Comparison with the ICE Conditions

This ICE Clause 15 is similar to the HKSAR Clause 17.

Comparison with the FIDIC Conditions

This FIDIC Clause 6.8 requires the Contractor to employ a sufficient number of persons having adequate knowledge of the language for communication and of the operations to be carried out. Obviously, for an international Contract, knowledge of the language is an important consideration. The Clause does not explicitly specify that the Contractor shall be represented full time by an authorized agent.

Clause 18

Contractor's Employees

(1) The Contractor shall provide and employ and shall ensure that any of his sub-contractors shall provide and employ on the Site in connection with the execution of the Works:

    (a) only such technical personnel as are skilled and experienced in their respective trades and callings and such sub-agents, foremen and leading hands as are competent to give proper supervision to the work they are required to supervise, and

    (b) such skilled, semi-skilled and unskilled labour as is necessary for the proper and timely execution of the Works.

(2) The Engineer shall be at liberty to object to and require the Contractor to remove forthwith from the Works any person employed by the Contractor or by a sub-contractor in or about the execution of the Works who in the opinion of the Engineer misconducts himself or is incompetent or negligent in the proper performance of his duties or fails to comply with any particular provision with regard to safety whose employment is otherwise considered by the Engineer to be undesirable and such person shall not be again employed upon the Works without the written permission of the Engineer.

(3) Any person so removed from the Works shall be replaced as soon as possible by a competent substitute.

Corresponding clauses in the ICE Conditions:

Clause 16

The Contractor shall employ or cause to be employed in the construction and completion of the Works and in the superintendence thereof only persons who are careful skilled and experienced in their several trades and callings.

    The Engineer shall be at liberty to object to and require the Contractor to remove or cause to be removed from the Works any person employed thereon who in the opinion of the Engineer misconducts himself or is incompetent or negligent in the performance of his duties or fails to conform with any particular provisions with regard to safety which may be set out in the Contract or persists in any conduct which is prejudicial to safety or health and such persons shall not be again employed upon the Works without the permission of the Engineer.

Corresponding clauses in the FIDIC Conditions:

Clause 6.9

The Contractor's Personnel shall be appropriately qualified, skilled and experienced in their respective trades or occupations. The Engineer may require the Contractor to remove (or cause to be removed) any person employed on the Site or Works, including the Contractor's Representative if applicable, who:

    (a) persists in any misconduct or lack of care,
    (b) carries out duties incompetently or negligently,
    (c) fails to conform with any provisions of the Contract, or
    (d) persists in any conduct which is prejudicial to safety, health, or the protection of the environment.

    If appropriate, the Contractor shall then appoint (or cause to be appointed) a suitable replacement person.

條款 18 「承建商」的僱員

(1) 「承建商」必須為有關「本工程」的實施，亦確保其「次承判商」為有關「本工程」的實施提供及僱用以下人士在「工地」工作：

   (a) 祇限於對他們各自的行業及職業有熟練技能和經驗的技術人員，及有能力妥善地監督所需工程的次代理人、管工及領班；及

   (b) 為妥善及按時地實施「本工程」所需要的熟練、半熟練及非熟練工人。

(2) 如「工程師」認為「承建商」或「次承判商」為實施「本工程」而僱用的任何人有不當行為或在妥善地履行職責方面不稱職或疏忽，或沒有依從某些關乎安全的條文，或「工程師」認為他不適宜受僱，「工程師」有權提出反對僱用此人，並要求「承建商」即時將此人遷出「本工程」。未經「工程師」書面許可，此人不可再受僱於「本工程」。

(3) 任何因此而被逐出「本工程」的人，須盡快由另一名稱職的代替人接替。

Commentary

This Clause requires the Contractor to employ only skilled and experienced technical personnel to plan and supervise the Works as well as skilled and semi-skilled labour for the proper and timely execution of the Works. Obviously, both quality and timely execution of the Works are important. Such requirements extend to those personnel to be provided by the sub-contractor. Misconduct, negligence and incompetence may result in any of the Contractor's or sub-contractor's personnel on Site being removed by the Engineer.

Analysis and application

This Clause 18 has a straightforward interpretation.

Comparison with the ICE Conditions

The ICE Clause is similar to the HKSAR Clause but additionally person failing to conform with safety provision or persists in conduct prejudicial to safety or health may be removed by the Engineer.

Comparison with the FIDIC Conditions

Apart from safety and health aspects, any conduct prejudicial to the protection of environment may also be a cause for removal.

Clause 20

The HKSAR Conditions:

Safety and Security of the Works

(1) The Contractor shall throughout the progress of the Works take full responsibility for the adequate stability and safety of all operations on the Site other than those of Specialist Contractors and utility undertakings and have full regard for the safety of all persons on the Site. The Contractor shall keep the Site and the Works in an orderly state appropriate to the avoidance of danger to all persons.

(2) The Contractor shall in connection with the Works provide and maintain all lights, guards, fences and warning signs and provide watchmen when and where necessary or required by the Engineer or by any competent statutory or other authority for the protection of the Works or for the safety and convenience of the public or others.

(3) The Contractor shall ensure that all parts of the Site where work is being carried out are so lighted as to ensure the safety of all persons on or in the vicinity of the Site and of such work.

(4) The Contractor, after obtaining any necessary approval from any relevant authority, shall submit to the Engineer proposals showing the layout of pedestrian routes, lighting, signing and guarding for any road opening or traffic diversion which may be required in connection with the execution of the Works. No such road opening or traffic diversion shall be brought into operation or use unless the proposals submitted have been previously approved by the Engineer and properly provided and implemented on the Site.

Corresponding clauses in the ICE Conditions:

Clause 19

(1) The Contractor shall throughout the progress of the Works have full regard for the safety of all persons entitled to be upon the Site and shall keep the Site (so far as the same is under his control) and the Works (so far as the same not completed or occupied by the Employer) in an orderly state appropriate to the avoidance of danger to such persons and shall among other things in connection with the Works provide and maintain at his own cost all lights guards fencing warning signs and watching when and where necessary or required by the Engineer's Representative or by any competent statutory or other authority for the protection of the Works or for the safety and convenience of the public or others.

Corresponding clauses in the FIDIC Conditions:

Clause 4.8

The Contractor shall:
(a) comply with all applicable safety regulations,
(b) take care for the safety of all persons entitled to be on the Site,
(c) use reasonable efforts to keep the Site and Works clear of unnecessary obstruction so as to avoid danger to these persons,
(d) provide fencing, lighting, guarding and watching of the Works until completion and taking over under Clause 10 [*Employer's Taking Over*], and
(e) provide any Temporary Works (including roadways, footways, guards and fences) which may be necessary, because of the execution of the Works, for the use and protection of the public and of owners and occupiers of adjacent land.

Clause 4.22

Unless otherwise stated in the Particular Conditions:
(a) the Contractor shall be responsible for keeping unauthorized persons off the Site, and
(b) authorised persons shall be limited to the Contractor's Personnel and the Employer's Personnel; and to any other personnel notified to the Contractor, by the Employer or the Engineer, as authorised personnel of the Employer's other contractors on the Site.

條款 20
「本工程」的安全及保護

(1) 在「本工程」施行過程中，除「專門承建商」及公用事業機構的操作外，「承建商」必須全面負責使「工地」內所有操作有足夠的穩定及安全，並必須全面照顧「工地」內所有的人的安全。「承建商」必須保持「工地」及「本工程」井然有序，以免對任何人構成危險。

(2) 「承建商」必須為保護「本工程」，或為公眾或其他人的安全和方便，或為「本工程」相關的事宜提供及維修所有照明、防護設施、圍欄和警告標誌。在有需要時或因應「工程師」或任何有法定權力的法定機關或其他主管當局的要求提供看守員保護「本工程」。

(3) 「承建商」須確保照亮「工地」內有工程施行的所有部分，以確保在「工地」內或其附近所有的人及此等工程的安全。

(4) 「承建商」從任何有關當局取得所需的准許後，必須向「工程師」提交建議，展示用於任何與「本工程」實施有關而可能需要的掘路工程或交通改道的行人路線、照明、標示及防護的鋪排。除非「承建商」所提交的建議事先已獲「工程師」批核及所建議的措施已在「工地」內妥善地提供及實施，否則「承建商」不可實施或作出此等掘路工程及交通改道。

Commentary

This Clause places an obligation on the Contractor to maintain safety on Site during the construction of the Works and he is responsible for the overall control of safety of all operations except those of the Employer's other contractors carrying out Specialist Works or utilities undertakers. The Contractor shall be responsible for maintaining lighting, guarding, fencing, warning signs and providing watchmen. He shall also maintain traffic and pedestrian routes and obtain approval from the relevant authorities, i.e., Transport Department, Police and Highways Department, etc., regarding his proposals. Further obligations on safety are usually given as a Special Condition and details of implementation are elaborated in the Particular Specification. For most major projects, the Contractor has to submit and implement a safety plan and employ safety officers.

Analysis and application

Clause 20(1) "… throughout the progress of the Works …"
The wording is unlikely to cover the Maintenance Period.

Clause 20(1) "… take full responsibility for the adequate stability and safety of all operations on the Site other than those of Specialist Contractor and utility undertakings …"
Specialist Contractor is defined in the HKSAR Conditions Clause 1 as any contractor employed by the Employer to execute Specialist Works. This Specialist Contractor must be distinguished from the Contractors on the Bureau's Approved List of Specialist Contractors and Suppliers in that the latter are employed by the Contractor and his acts are still the responsibility of the Contractor.

Clause 20(2) "… provide and maintain all lights, guards, fences and warning signs and provide watchmen …"
If any accident occurs on the Site, the victim may sue both the Contractor and the Employer for negligence and therefore it is always in the Contractor's interests to keep the Site safe at any hours of the day. Under Clause 22, the Contractor shall indemnify the Employer against all loss and claims for injury or damage to any person or property, and therefore the Employer shall simply pass all claims arising from injury or damage to the Contractor under the indemnity clause.

Comparison with the ICE Conditions

This ICE Clause 19(1) is very similar to the HKSAR Clause 20. As long as the control of the Site remains in the hands of the Contractor and the Works are still being carried out by the Contractor, he shall have full regards for the safety of all persons entitled to be upon the Site and keep the Site in an orderly state. For those persons not authorized to be on Site, e.g. trespassers, the Contractor as an occupier still has some duty of common humanity or taking reasonable care. Putting up warning signs are always appropriate but special care would be needed if the Site is near a place frequented by children, such as a school.

Comparison with the FIDIC Conditions

The FIDIC Clause 4.8 is similar to the HKSAR Clause 20 and the ICE Clause 19(1). In addition, the FIDIC Clause 4.22 requires the Contractor to keep unauthorized persons off the Site. The Contractor may have to provide guards and watchmen to achieve this requirement.

 10 Can you distinguish which parts of these falsework are Constructional Plant and which parts are Temporary Works? Are these falsework supplied by the Specialist Contractor?

The Contractor is now providing much more guarding and lighting to the public road and fencing to the private developments than he has envisaged at the Tender. Can the Contractor make a claim? How about under the ICE Conditions or FIDIC Conditions?

Clause 21

Care of the Works

(1) From and including the date for commencement of the Works notified by the Engineer in accordance with Clause 47 until 28 days after the date of completion of the Works certified by the Engineer in accordance with Clause 53, or until the date the Employer takes over the Works, if earlier, the Contractor shall take full responsibility for the care of the Works and any Specialist Works (except the stability and safety of the operations of Specialist Contractors referred to in Clause 20(1)) or any part thereof, and for the care of any Constructional Plant, temporary buildings and materials and things whatsoever on the Site or delivered to or placed on the Site in connection with or for the purpose of the Works or any Specialist Works.

Provided that if the Engineer shall issue a certificate of completion in respect of any Section or part of the Works before he shall issue a certificate of completion in respect of the Works the Contractor shall cease to be responsible for the care of that Section or part of the Works 28 days after the date of completion certified by the Engineer in respect of that Section or part and the responsibility for the care thereof shall thereupon pass to the Employer.

Provided further that the Contractor shall take full responsibility for the care of any outstanding work which he shall have undertaken to finish during the Maintenance Period until such outstanding work is complete, and shall continue to be responsible for all things which are required to be retained on the Site during the Maintenance Period including Constructional Plant, temporary buildings and materials and other facilities provided for the use of the Engineer, the Engineer's Representative and their staff.

Corresponding clauses in the ICE Conditions:

Clause 20

(1) (a) The Contractor shall save as in paragraph (b) hereof and subject to sub-clause (2) of this Clause take full responsibility for the care of the Works and materials plant and equipment for incorporation therein from the Works Commencement Date until the date of issue of a Certificate of Substantial Completion for the whole of the Works when the responsibility for the said care shall pass to the Employer.

(b) If the Engineer issues a Certificate of Substantial Completion for any Section or part of the Permanent Works the Contractor shall cease to be responsible for the care of that Section or part from the date of issue of that Certificate of Substantial Completion when the responsibility for the care of that Section or part shall pass to the Employer.

(c) The Contractor shall take full responsibility for the care of any work and materials plant and equipment for incorporation therein which he undertakes during the Defects Correction Period until such work has been completed.

Corresponding clauses in the FIDIC Conditions:

Clause 17.2 (*part*)

The Contractor shall take full responsibility for the care of the Works and Goods from the Commencement Date until the Taking-Over Certificate is issued (or is deemed to be issued under Sub-Clause 10.1 [*Taking Over the Works and Sections*]) for the Works, when responsibility for the care of the Works shall pass to the Employer. If a Taking-Over Certificate is issued (or is so deemed to be issued) for any Section or part of the Works, responsibility for the care of the Section or part shall then pass to the Employer.

After responsibility has accordingly passed to the Employer, the Contractor shall take responsibility for the care of any work which is outstanding on the date stated in a Taking-Over Certificate, until this outstanding work has been completed.

條款 21

對「本工程」的照顧

(1) 從「工程師」根據第47條規定通知的「本工程」開工日期起（包括該日），直到「工程師」根據第53條核實為「本工程」的竣工日期後28天或「僱主」接管「本工程」的日期（以早者為準），「承建商」對「本工程」或其任何部份及任何「專門工程」的照顧須全面負責（第20條(1)款所提及的「專門承建商」及公用事業機構操作的穩定及安全除外），亦必須對與「本工程」或「專門工程」有關的或為其實施目的的任何「施工設備」、臨時建築物、物料及送往或放於工地的任何物品的照顧全面負責。

但如「工程師」為「本工程」頒發竣工證明書之前頒發「本工程」的任何「工段」或部分竣工證明書，則由「工程師」核實為此「工段」或此部分的竣工日期後28天起，「承建商」將不會對照顧此「工段」或該部分負責。由此時起，照顧的責任將轉交「僱主」。

但再者，對於「承建商」未完成而已承諾必須於「保養期」內完成的工程，「承建商」必須對其照顧全面負責，直至此等未完成的工程完成為止。「承建商」亦必須對有需要在「保養期」內保留於「工地」的一切物品，包括「施工設備」、臨時建築物料及其他提供「工程師」、「工程師代表」及其職員使用的設施繼續負責。

Commentary

This Clause 21 is one of, if not the most, important clause in the HKSAR Conditions since it defines the sharing of construction risks between the Employer and the Contractor and the exact times when the construction risks of the Works are passed back to the Employer. For example, when a heavy storm damages the Works during the construction of the Works, the Contractor has to be responsible for the damage that has actually occurred even if the Contractor has taken all necessary precaution against any damage and no matter how unforeseeable is the consequence of the heavy storm. However, if the heavy storm comes beyond 28 days after the Engineer has certified completion of the Works or the Employer has taken over the Works, any damage to the Works by the heavy storm becomes the responsibility of the Employer. Apart from the timing of the transfer of the risks, the clause also defines the types of risks which are not the responsibility of the Contractor and which are retained by the Employer. These are called "excepted risks" as defined in sub-clause (4), such as war, defective design by the Engineer or risk arising from the default of the Employer or Engineer. It is to be noted that, in addition to the Works, the Contractor also takes responsibility for the Specialist Works constructed by the Employer's other contractors as well as the Constructional Plant. In case of any damage or loss to the Works, Specialist Works or Constructional Plant, the Contractor shall be obliged to speedily make good such loss or damage, or at the option of the Employer, pay the Employer for making good such loss or damage. During the Maintenance Period, there may still be a lot of outstanding works to be completed, and the Contractor shall take responsibility of these outstanding works and all the things required to be retained on Site until the outstanding works are completed. In this connection, the Contractor has to be responsible for the care of all the temporary offices, including the Engineer's site office, until these are demolished and cleared, usually after the Maintenance Period.

Analysis and application

Clause 21(1) "From and including the date for commencement of the Works notified by the Engineer … until 28 days after the date of completion of the Works certified by the Engineer in accordance with Clause 53 …"

The 28 days are normally intended for the project department to hand over the respective parts of the Works to the various maintenance authorities such as the Highways Department, Water Supplies Department or Electrical and Mechanical Department. However, a difficulty often arises because the Engineer seldom issues the certificate of completion on the date of completion of the Works but usually backdates the completion to an earlier date which may be several weeks or even months before the issue of the certificate of completion. The Engineer allows himself to finalize any extension of time, which he may grant to the Contractor, obviating the premature imposition of liquidated damages. This sometimes creates problems regarding insurance in that loss or damage may occur during the period when the date of completion still has to be determined by the Engineer. It is uncertain whether the responsibility for the care of works lies with the Contractor or the Employer. The Engineer has to deal with such situations very carefully and especially make sure that there is insurance cover for the questionable period.

Clause 21(1) "... until the date the Employer takes over the Works, if earlier ..."

It is not uncommon that the taking over process may take a period of several weeks to several months and as explained in the preceding paragraph, the Engineer needs to make sure that the Works are covered by insurance, if necessary.

Clause 21(1) "... the Contractor shall take full responsibility for the care of the Works and any Specialist Works (except the stability and safety of the operation of the Specialist Contractors referred to in Clause 20(1)) ..."

Before certifying completion of the Works, the Engineer has to see that the Works are completed to his satisfaction and the Contractor has to undertake the completion of any outstanding works. The above Clause reaffirms the Contractor's liability to make good any damage to the Works to the satisfaction of the Engineer. Regarding Specialist Works carried out by the Employer's other contractors, the wording in the brackets excludes the Contractor's responsibility for the stability and safety of the operations of those contractors. The word "operations" implies that during construction, the Specialist Contractors are responsible but the responsibility is passed to the Contractor after the completion of the Specialist Works.

Clause 21(1) "... and for the care of any Constructional Plant, temporary buildings and materials and things whatsoever on the Site or delivered to or placed on the Site in connection with or for the purpose of the Works or any Specialist Works ..."

It should be noted that the Contractor is even responsible for the care of the Constructional Plant, temporary buildings and materials of the Specialist Contractors. However, it must be noted that according to the definition, all Specialist Works must have been "separately identified in the Contract," which means that for any Specialist Works added during the construction period, the Employer may have to pay extra for their care.

Clause 21(1) "... Provided that if the Engineer shall issue a certificate of completion in respect of any Section or part of the Works ..."

That means all the above discussions apply similarly to any Section or part of the Works.

Clause 21(1) "... Provided further that the Contractor shall take full responsibility for the care of any outstanding work ..."

The position regarding the care of the Works is uncertain when the Engineer certifies completion but with substantial outstanding work. For example, a slope is completed apart from a part of slope protection work and the Engineer certifies the slope to be substantially completed. When a heavy storm causes a major collapse, the Contractor would argue that he is only responsible for the care of the part of slope protection work and the Employer has to bear the consequence of failure. Sometimes, this may occur before the work is handed over to the maintenance authority and there may be a gap in time where neither the Contractor nor the maintenance authority could be responsible. Therefore, the Engineer has to consider very carefully when drawing up the list of outstanding work before certifying completion.

Q 12 This is a tower for the cable-stayed bridge and is found on heavy piling. It will have a height of more than 200 metres high upon completion. The cost is in excess of $200M. Can the Contractor apply for partial completion after it is completed?

Clause 21

Care of the Works

(continued)

(2) In case any damage, loss or injury from any cause whatsoever, except the "excepted risks" as defined in sub-clause (4) of this Clause, shall happen to the Works or any Specialist Works or any part thereof, or to any Constructional Plant, temporary buildings, materials and things whatsoever on the Site, the Contractor shall at his own expense and with all possible speed make good or at the option of the Employer shall pay to the Employer the cost of making good any such damage, loss or injury to the satisfaction of the Engineer and shall, notwithstanding such damage, loss or injury, proceed with the execution of the Works in all respects in accordance with the Contract and the Engineer's instructions.

(3) To the extent that any damage, loss or injury arises from any of the "excepted risks" defined in sub-clause (4) of this Clause, the Contractor shall, if instructed by the Engineer, repair and make good the same at the expense or proportionate expense of the Employer. Any sum payable under this Clause by the Employer shall be valued by the Engineer in the same manner as a sum payable in respect of a variation ordered in accordance with Clause 60.

Corresponding clauses in the ICE Conditions:

Clause 20

(3) (a) In the event of any loss or damage to
   (i) the Works or any Section or part thereof or
   (ii) materials plant or equipment for incorporation therein while the Contractor is responsible for the care thereof (except as provided in sub-clause (2) of this Clause) the Contractor shall at his own cost rectify such loss or damage so that the Permanent Works conform in every respect with the provisions of the Contract and the Engineer's instructions. The Contractor shall also be liable for any loss or damage to the Works occasioned by him in the course of any operation carried out by him for the purpose of complying with his obligations under Clauses 49 and 50.

(b) Should any such loss or damage arise from any of the Excepted Risks defined in sub-clause (2) of this Clause the Contractor shall if and to the extent required by the Engineer rectify the loss or damage at the expense of the Employer.

(c) In the event of loss or damage arising from an Excepted Risk and a risk for which the Contractor is responsible under sub-clause (1)(a) of this Clause then the Engineer shall when determining the expense to be borne by the Employer under the Contract apportion the cost of rectification into that part caused by the Excepted Risk and that part which is the responsibility of the Contractor.

Corresponding clauses in the FIDIC Conditions:

Clause 17.2 (*continued*)

If any loss or damage happens to the Works, Goods or Contractor's Documents during the period when the Contractor is responsible for their care, from any cause not listed in Sub-Clause 17.3 [*Employer's Risks*], the Contractor shall rectify the loss or damage at the Contractor's risk and cost, so that the Works, Goods and Contractor's Documents conform with the Contract.

The Contractor shall be liable for any loss or damage caused by any actions performed by the Contractor after a Taking-Over Certificate has been issued. The Contractor shall also be liable for any loss or damage which occurs after a Taking-Over Certificate has been issued and which arose from a previous event for which the Contractor was liable.

Clause 17.4

If and to the extent that any of the risks listed in Sub-Clause 17.3 above results in loss or damage to the Works, Goods or Contractor's Documents, the Contractor shall promptly give notice to the Engineer and shall rectify this loss or damage to the extent required by the Engineer.

If the Contractor suffers delay and/or incurs Cost from rectifying this loss or damage, the Contractor shall give a further notice to the Engineer and shall be entitled subject to Sub-Clause 20.1 [*Contractor's Claims*] to:
(a) an extension of time for any such delay, if completion is or will be delayed, under Sub-Clause 8.4 [*Extension of Time for Completion*], and
(b) payment of any such Cost, which shall be included in the Contract Price. In the case of sub-paragraphs (f) and

(g) of Sub-Clause 17.3 [*Employer's Risks*], reasonable profit on the Cost shall also be included.

After receiving this further notice, the Engineer shall proceed in accordance with Sub-Clause 3.5 [*Determinations*] to agree or determine these matters.

條款 21

香港特別行政區土木工程合同一般條件：

對「本工程」的照顧

（續）

(2) 除本條第(4)款訂明的「豁免風險」外，如「本工程」或任何「專門工程」或其任何部分、或「工地」內任何「施工設備」、臨時建築、物料及任何物品，無論出於任何原因而受任何損害、損失或人身損傷，「承建商」必須自費及以盡快速度將此等損害、損失或人身損傷回復至「工程師」滿意為止，或按照「僱主」的選擇，向「僱主」支付將其回復至「工程師」滿意為止的費用。雖然出現此等損害、損失或人身損傷，「承建商」仍然必須在各方面按照「合同」及「工程師」的指令繼續實施「本工程」。

(3) 在任何因為本條第(4)款訂明的「豁免風險」而引起的損害、損失或人身損傷的範圍內，「承建商」必須在接到「工程師」的指令後將此等損害、損失或人身損傷修理及復原，此費用由「僱主」支付或按比例支付。任何根據本條必須由「僱主」支付的金額必須經「工程師」釐訂，釐訂方法須等同根據第60條內對作出工程變更的命令須支付的金額的釐訂方法進行。

Analysis and application (continued)

Clause 21(2) "In case any damage, loss or injury from any cause whatsoever, except the 'excepted risks' … the Contractor shall at his own expense and with all possible speed make good or at the option of the Employer shall pay to the Employer the cost of making good any such damage, loss or injury …"

By this sub-clause, the Contractor has to bear all the cost of making good the damage and he has to carry out the repair work with all possible speed in order that the overall progress of the Works will not be seriously affected. The wording "at the option of the Employer" suggests that the Employer can choose to use his own workmen to make good any damage without waiting for the Contractor to do so and the Contractor must pay the Employer for it or the Employer may deduct from monies due to the Contractor from the interim certificate or retention money in accordance with Clause 83. This right of the Employer is apparently not dependent on the right under Clauses 82(1) and 82(2), which could only be invoked if the Contractor fails, or is unable or unwilling to carry out the repair works.

Clause 21(3) "To the extent that any damage, loss or injury arises from any of the 'excepted risks' defined in sub-clause (4) of this Clause, the Contractor shall, if instructed by the Engineer, repair and make good the same at the expense or proportional expense of the Employer …"

This sub-clause allows the Engineer to instruct the Contractor to make good damage even the damage arises from a cause which the Employer is responsible, e.g., design fault. If the damage arises only partly from a cause which the Employer is responsible, the sub-clause allows for the apportionment of cost between the Employer and the Contractor.

The HKSAR Conditions:

Care of the Works

(4) The "excepted risks" are:

(a) outbreak of war (whether war be declared or not) in which Hong Kong shall be actively engaged;

(b) invasion of Hong Kong;

(c) act of foreign terrorists in Hong Kong;

(d) civil war, rebellion, revolution or military or usurped power in Hong Kong;

(e) riot, commotion or disorder in Hong Kong otherwise than amongst the employees of the Contractor, any sub-contractor or Specialist Contractor currently or formerly engaged on the Works or Specialist Works;

(f) a cause due to the occupation by the Employer, his agents, employees or other contractors of any part of the Works for a purpose other than carrying out of Specialist Works, such purpose being authorized and required by the Employer;

(g) damage, loss or injury which is the direct consequence of the Engineer's design of the Works;

(h) a cause due to any neglect or default by the Engineer or the Employer or their employees or agents in the course of their employment;

(i) ionising radiations or contamination by radio-activity from any nuclear fuel or from any nuclear waste from the combustion of nuclear fuel, radio-active toxic explosive, or other hazardous properties of any explosive nuclear assembly or nuclear component thereof provided always that the same are not caused in whole or in part by the Contractor or any sub-contractor.

Corresponding clauses in the ICE Conditions:

Clause 20 (*continued*)

(2) The Excepted Risks for which the Contractor is not liable are loss or damage to the extent that it is due to

(a) the use or occupation by the Employer his agents servants or other contractors (not being employed by the Contractor) of any part of the Permanent Works

(b) any fault defect error or omission in the design of the Works (other than a design provided by the Contractor pursuant to his obligations under the Contract)

(c) riot war invasion act of foreign enemies or hostilities (whether war be declared or not)

(d) civil war rebellion revolution insurrection or military or usurped power

(e) ionizing radiations or contamination by radioactivity from any nuclear fuel or from any nuclear waste from the combustion of nuclear fuel radioactive toxic explosive or other hazardous properties of any explosive nuclear assembly or nuclear component thereof and

(f) pressure waves cause by aircraft or other aerial devices travelling at sonic or supersonic speeds.

Corresponding clauses in the FIDIC Conditions:

Clause 17.3

The risks referred to in Sub-Clause 17.4 below are:

(a) war, hostilities (whether war be declared or not), invasion, act of foreign enemies,

(b) rebellion, terrorism, revolution, insurrection, military or usurped power, or civil war, within the Country,

(c) riot, commotion or disorder within the Country by persons other than the Contractor's Personnel and other employees of the Contractor and Subcontractors,

(d) munitions of war, explosive materials, ionizing radiation or contamination by radio-activity, within the Country, except as may be attributable to the Contractor's use of such munitions, explosives, radiation or radio-activity,

(e) pressure waves caused by aircraft or other aerial devices travelling at sonic or supersonic speeds,

(f) use or occupation by the Employer of any part of the Permanent Works, except as may be specified in the Contract,

(g) design of any part of the Works by the Employer's Personnel or by others for whom the Employer is responsible, and

(h) any operation of the forces of nature which is unforeseeable or against which an experienced contractor could not reasonably have been expected to have taken adequate preventative precautions.

對「本工程」的照顧

(4)　「豁免風險」指：

　(a)　香港積極參與的戰爭（無論有宣戰與否）；

　(b)　香港受到侵襲；

　(c)　外地恐怖分子在香港的活動；

　(d)　香港發生內戰、暴動、革命，或出現軍事政變或篡奪政權；

　(e)　香港發生暴動、騷動或混亂，但對局限在目前或以前從事「本工程」或「專門工程」的「承建商」、任何「次承判商」或「專門承建商」的僱員之間的此類情況除外；

　(f)　除為實施「專門工程」外，「僱主」、「僱主」代理人、僱員或其他承建商對「本工程」任何部分的佔用，而此佔用是由「僱主」授權和要求的；

　(g)　由於「工程師」的工程設計直接引致的損害、損失或人身損傷；

　(h)　由於「工程師」或「僱主」或他們的僱員或代理人在受僱期間的疏忽或失責引致的事件；

　(i)　從任何核燃料，或從核燃料燃燒而產生的核廢料、放射性或有毒炸藥，或從任何爆炸性的核組件或部件的危險特質所引致的輻射或放射性污染，而該污染並非完全或部分由「承建商」或任何「次承判商」所引致的。

Analysis and application (continued)

Clause 21(4) "The 'excepted risks' …"

The excepted risks, such as the outbreak of war, invasion of Hong Kong, etc., from (a) to (e) are the risks which are usually excluded in any standard construction-all-risks insurance. For item (f) regarding occupation by the Employer, it may be required before or after the Works or part of the Works is completed and the Employer would be responsible for the care of that relevant part of the Works. Again, if the relevant requirement has not been included in the Contract, the insurance may not cover such risks and the Employer has to check its insurance position. Item (g) is the design fault which are usually covered by insurance. Item (h) concerns damage caused by the neglect or default of the Engineer or the Employer or their employees or agents in the course of their employment. The Employer's default would probably be covered by standard insurance policy if the Works are insured in the joint name of the Employer and Contractor. The risk arising from the Engineer's default is usually covered by the professional indemnity insurance. It has to be noted that even if the risks are covered by the insurance, the Party responsible for the risk has to pay the deductibles or excess set by the insurance company. There appears to be a contradiction between Clause 2(4) which states that no act or omission in the performance of duty by the Engineer shall relieve the Contractor of his obligations and Clause 21(h) which absolves the Contractor's responsibility for the care of the Works if the damage arises from neglect or default by the Engineer. It is submitted that Clause 2(4) only applies if the Engineer's act is not positively contravening the provisions of the Contract.

Comparison with the ICE Conditions

This ICE Clause 20 is similar to the HKSAR Clause. One major difference is that the responsibility for the care of the Works will only pass after "the date of issue of a Certificate of Substantial Completion," which removes any uncertainty in the HKSAR Conditions when the Engineer has to backdate the date of completion to an earlier date. The Engineer would not issue the Certificate of Completion until the Works are ready for taking over by the maintenance authority. There would not be a gap where neither the Employer nor the Contractor is both not responsible for the care of the Works. Another difference is that the ICE Clause does not provide an option for the Employer to carry out the repair works by his own workmen. The Employer has to rely on the ICE Clause 62 to give him power to employ his own workmen to carry out the repair works if the Contractor is "unable or unwilling" to do so. The last difference is that the "excepted risks" in the ICE Conditions do not include "any neglect or default by the Engineer or Employer," probably because the ICE Conditions require compulsory construction-all-risks insurance issued in the joint name of the Employer and the Contractor to make arrangement to cover such risks.[1]

Comparison with the FIDIC Conditions

This FIDIC Clause 17 is very similar to the ICE Clause 20 with the same differences as compared to the HKSAR Clause. In addition, it also specifically mentions the responsibility of care of Goods and Contractor's Documents. The Contractor is also liable for any loss or damage, which occurs after a Taking-Over Certificate has been issued, that arises from a previous event which the Contractor was liable.

Notes

1. Max Abrahamson, *Engineering Law and the ICE Contracts*, 4th Edition, p. 97.

 The Government Stonecutter Sewage Treatment Work is in close proximity to the massive flyover construction. Who is responsible if the Contractor accidentally damages the Sewage Treatment Works during the flyover construction?

Can you suggest all the necessary insurance policies that should be taken out by the Contractor for this heavy cable-stayed construction?

<table>
<tr>
<td>

Clause

22

</td>
<td>

The HKSAR Conditions:

Damage to Persons and Property

</td>
</tr>
</table>

(1) The Contractor shall, except if and so far as the Contract otherwise provides, indemnify and keep indemnified the Employer against all losses and claims for injury or damage to any person or property whatsoever, other than surface or other damage to land or crops on the Site, which may arise out of or in consequence of the execution of the Works and against all claims, demands, proceedings, damages, costs, charges and expenses whatsoever in respect thereof or in relation thereto.

(2) The Contractor shall make good or at the option of the Employer shall pay to the Employer the cost of making good any damage, loss or injury which may occur to any property of the Employer and shall recompense the Employer in respect of any damage, loss or injury which may occur to any agent or employee of the Employer by or arising out of or in consequence of the execution of the Works or in the carrying out of the Contract.

Corresponding clauses in the ICE Conditions:

Clause 22

(1) The Contractor shall except if and so far as the Contract provides otherwise and subject to the exceptions set out in sub-clause (2) of this Clause indemnify and keep indemnified the Employer against all losses and claims in respect of

   (a) death of or injury to any person or

(b) loss of or damage to any property (other than the Works)

which may arise out of or in consequence of the construction of the Works and the remedying of any defects therein and against all claims demands proceedings damages costs charges and expenses whatsoever in respect thereof or in relation thereto.

Corresponding clauses in the FIDIC Conditions:

Clause 17.1

The Contractor shall indemnify and hold harmless the Employer, the Employer's Personnel, and their respective agents, against and from all claims, damages, losses and expenses (including legal fees and expenses) in respect of:

(a) bodily injury, sickness, disease or death, of any person whatsoever arising out of or in the course of or by reason of the Contractor's design (if any), the execution and completion of the Works and the remedying of any defects, unless attributable to any negligence, wilful act or breach of the Contract by the Employer, the Employer's Personnel, or any of their respective agents, and

(b) damage to or loss of any property, real or personal (other than the Works), to the extent that such damage or loss:

   (i) arises out of or in the course of or by reason of the Contractor's design (if any), the execution and completion of the Works and the remedying of any defects, and

   (ii) is attributable to any negligence, wilful act or breach of the Contract by the Contractor, the Contractor's Personnel, their respective agents, or anyone directly or indirectly employed by any of them.

條款 22

對人及財產的損害

(1) 除「合同」另有規定外並限於其所載情況下，「承建商」必須保障「僱主」免除因實施「本工程」而可能引起或造成的一切損失及對任何人受損傷或財產受損害（「工地」內的土地面層或其他土地或農作物損害除外）的索償，及一切因此所牽涉或與此有關的任何索償、要求、訴訟、賠償、訟費、收費及開支，並使「僱主」繼續可得到此保障。

(2) 對於「承建商」實施「本工程」或履行「合同」所引致或造成「僱主」的任何財產受到的損害、損失或人身損傷，「承建商」必須使其復原或按「僱主」的選擇向「僱主」支付使其復原的費用。對「僱主」的任何代理人或僱員所受到的任何損害、損失或人身損傷，「承建商」也必須向「僱主」作出補償。

Commentary

This Clause 22 generally provides indemnity to the Employer against any claims for injury or damage by the third party against the Employer as a result of construction works carried out by the Contractor, subject to the exceptions described in sub-clauses (2)(b)(i) and (ii) regarding the use or occupation by the Employer or any interference of others' right which is the unavoidable results of executing the Works in accordance with the Contract. The Clause also provides for the apportionment of costs arising from damage or injury between the Employer and the Contractor if the damage or injury is caused partly by the default of the Engineer or the Employer. Sub-clause (2) obliges the Contractor to make good any damage to the Employer's property or to pay to the Employer if the Employer wishes to carry out the repair work himself.

Analysis and application

Clause 22(1) "The Contractor shall, … indemnify and keep indemnified the Employer against all losses and claims for injury or damage to any person or property whatsoever, other than surface or other damage to land or crops on the Site, which may arise out of or in consequence of the execution of the Works …"

"To indemnify the Employer against all losses" means to protect the Employer from any financial consequence resulting from any injury or damage to the third party. When the Employer receives any notice of claim from the third party, he can simply pass the notice to the Contractor who will deal with the claim and if legal proceedings are instituted against the Employer and in many cases also against the Contractor and the Engineer, the Employer can rely on the Contractor to employ lawyer to defend the case. However, if the Employer, for political or other reasons wishes, to have his own legal representation, he may not arguably get back his legal cost from the Contractor. In many cases, the Employer may like to see that the Contractor settles the case with the claimant as soon as possible but he should refrain from any direct involvement in the settlement process as far as possible since the Contractor is entitled to defend his case and there is also complication regarding the insurance position. The wording "other than surface or other damage to land … on the Site" usually means that the Contractor will not be responsible for any damage done if the Employer fails to obtain a legal right to construct the works over or through the land. The damage can be caused by both the permanent works and all necessary Temporary Works. It is therefore essential that the Engineer and the Employer clear all the land issues before calling the Tender. The wording "damage to … crops" may include fruit trees or trees for commercial use.

Clause 22(2) "The Contractor shall make good or at the option of the Employer shall pay to the Employer the cost of making good any damage, loss or injury which may occur to any property of the Employer …"

This sub-clause requires the Contractor to make good any damage caused to the Employer's property, e.g., for a roadwork contract, the Employer's property includes the existing road and road furniture. In many situations, the Employer would like its maintenance authority to repair the damage and recover payment from the Contractor for the work and this sub-clause allows the Employer to have the option to do so.

Clause 22(2) "… and shall recompense the Employer in respect of any damage, loss or injury which may occur to any agent or employee of the Employer …"

The Government of the HKSAR does not usually take out insurance for its employees and it can rely on this sub-clause to cover any loss arising from injury to the employee.

Q 15 What is the Contractor's position if during the typhoon, some container boxes are blown off into the heavy falsework system (in the foreground) causing substantial damage?

 Q 16 What is the contractual position if the tower crane for the concrete tower construction falls onto the adjacent (in the foreground) container operator's land? What is the usual insurance position?

Clause 22

Damage to Persons and Property

(continued)

(2) Provided that:

(a) the Contractor's liability to indemnify or recompense the Employer under sub-clauses (1) and (2) of this Clause shall, subject to sub-clause (3) of this Clause, be reduced proportionately to the extent that the act or neglect of the Engineer or the Employer, their respective agents or employees shall have contributed to the damage, loss or injury;

(b) nothing herein contained shall be deemed to render the Contractor liable for or in respect of or to indemnify the Employer against any compensation or damages for or with respect to:

(i) the use or occupation of land provided by the Employer for the Works, or for the purpose of executing the Works, or interference, whether temporary or permanent, with any right of way, light, air or water or other easement or quasi easement which is the unavoidable result of the execution of the Works in accordance with the Contract,

(ii) the right of the Employer to construct the Works on, over, under, in or through any land, or for or in respect of all claims, demands, proceedings, damages, costs, charges and expenses in respect thereof or in relation thereto.

(3) The indemnities given herein by the Contractor shall not be rendered ineffective or reduced by reason of any negligence or omission of the Employer or the Engineer or the Engineer's Representative in watching and inspecting the Works, or in testing and examining any material to be used and workmanship employed by the Contractor in connection with the Works, or in supervising or controlling the Contractor's site operations or methods of working or Temporary Works, or in detecting or preventing or remedying defective work or services, or in ensuring proper performance of any other obligation of the Contractor.

Corresponding clauses in the ICE Conditions:

Clause 22 (*continued*)

(2) The exceptions referred to in sub-clause (1) of this Clause which are the responsibility of the Employer are

(a) damage to crops being on Site (save in so far as possession has not been given to the Contractor)

(b) the use or occupation of land provided by the Employer for the purposes of the Contract (including consequent losses of crops) or interference whether temporary or permanent with any right of way light air or water or other easement or quasi-easement which are the unavoidable result of the construction of the Works in accordance with the Contract

(c) the right of the Employer to construct the Works or any part thereof on over under in or through and land

(d) damage which is the unavoidable result of the construction of the Works in accordance with the Contract and

(e) death of or injury to persons or loss of or damage to property resulting from any act neglect or breach of statutory duty done or committed by the Employer his agents servants or other contractors (not being

employed by the Contractor) or for or in respect of any claims demands proceedings damages costs charges and expenses in respect thereof or in relation thereto.

(3) The Employer shall subject to sub-clause (4) of this Clause indemnify the Contractor against all claims demands proceedings damages costs charges and expenses in respect of the matters referred to in the exceptions defined in sub-clause (2) of this Clause.

(4) (a) The Contractor's liability to indemnify the Employer under sub-clause (1) of this Clause shall be reduced in proportion to the extent that the act or neglect of the Employer his agents servants or other contractors (not being employed by the Contractor) may have contributed to the said death injury loss or damage.

(b) The Employer's liability to indemnify the Contractor under sub-clause (3) of this Clause in respect of matters referred to in sub-clause 2(e) of this Clause shall be reduced in proportion to the extent that the act or neglect of the Contractor or his sub-contractors servants or agents may have contributed to the said death injury loss or damage.

Corresponding clauses in the FIDIC Conditions:

Clause 17.1 (*continued*)

The Employer shall indemnify and hold harmless the Contractor, the Contractor's Personnel, and their respective agents, against and from all claims, damages, losses and expenses (including legal fees and expenses) in respect of (1) bodily injury, sickness, disease or death, which is attributable to any negligence, wilful act or breach of the Contract by the Employer, the Employer's Personnel, or any of their respective agents, and (2) the matters for which liability may be excluded from insurance cover, as described in sub-paragraphs (d)(i), (ii) and (iii) of Sub-Clause 18.3 [*Insurance Against Injury to Persons and Damage to Property*].

條款 22

香港特別行政區土木工程合同一般條件：

對人及財產的損害

（續）

(2) 但是：

(a) 在符合本條第(3)款規定下，「承建商」根據本條第(1)及(2)款要保障「僱主」免除索償的法律責任，須視乎「工程師」或「僱主」、及其各自的代理人或僱員的行為或疏忽所引致的損害、損失或人身傷害應分擔責任的程度，相應按比例減少；

(b) 本條所載的規定不可被視為使「承建商」因為或就下列事宜而有法律責任作出任何補償或賠償或為此等補償或賠償而保障「僱主」免除索償：

(i) 「承建商」使用或佔用「僱主」為「本工程」或為其實施目的而提供的土地，或「承建商」對任何通行權、光線、空氣或水、或其他地役權或類似地役權的干擾，無論是暫時或永久的，而這些干擾是因為「承建商」按照「合同」實施「本工程」而無可避免的；

(ii) 「僱主」在任何土地、其上面、其下面、其內面或穿過該土地實施「本工程」的權利，或因為或就上文所述而牽涉或與其有關的一切索償、要求、訴訟、賠償、訟費、收費及開支。

(3) 按照本條文由「承建商」作出的保障，不可因為「僱主」或「工程師」或「工程師代表」對「本工程」的監察及視察，或對「承建商」為有關「本工程」使用的任何物料及採用的技術水平而進行的測試及檢查，或對「承建商」的工地操作或工作方法或其「臨時工程」的監督或控制，或對欠妥的工程或服務的測查或防止或修補，或確保「承建商」的其他責任可以妥善地履行的任何方面的疏忽或不作為而變成無效或因此而減少。

Analysis and application (continued)

Clause 22(2)(a) "The Contractor's liability … be reduced proportionately to the extent that the act or neglect of the Engineer or the Employer … shall have contributed to the damage, loss or injury;"

This allows for reduction of Contractor's liability arising from the act or neglect of the Engineer or the Employer. However, the HKSAR Conditions sub-clause 2(4) states that "No act or omission by the Engineer … in the performance of his duties … shall in any way operate to relieve the Contractor of any of his duties, responsibilities, obligations or liabilities …" It is submitted that sub-clause 2(4) applies if the Engineer acts generally in accordance with the Contract but there may be some carelessness or omission in carrying out his duties especially his supervisory role. However, if the Engineer's act is positively contrary to the provisions of the Contract and cause third party damage, then the act or neglect of the Engineer will obviate or reduce the Contractor's liability.

Clause 22(2)(b) "nothing therein contained shall be deemed to render the Contractor liable ... (i) the use or occupation of land ... or interference, whether temporary or permanent, with any right of way ... or other easement or quasi easement which is the unavoidable result of the execution of the Works in accordance with the Contract."

The wording "the use or occupation of land" covers damage due to the long-term effect of the completed Works on the claimant's interests. The wording "interference, whether temporary or permanent, with any right of way ..." covers roads, natural light, air or waterway of both permanent and temporary nature. The wording "other easement" may include right of support to building, etc. The wording "quasi easement" may include restrictive covenants and local customary rights. The wording "unavoidable result of the execution of the Works in accordance with the Contract" implies that the Contractor must execute the Works with minimum or at most reasonable working space.

Clause 22(2)(b)(ii) "the right of the Employer to construct the Works on, over, under, in or through any land, ..."

Any action of trespass may be brought against the Contractor if, for example, some soil nails or anchors encroach onto the private land and therefore the Contractor has to be indemnified by the Employer.

Clause 22(3) "The indemnities given herein by the Contractor shall not be rendered ineffective or reduced by reason of any negligence or omission of the Employer or the Engineer or the Engineer's Representative in watching and inspecting the Works, ... or in supervisory or controlling the Contractor's site operations ..."

This means that the liability of the Contractor would not be rendered ineffective or reduced on account of the Employer's or Engineer's failure to supervise or control the Contractor's Site activities and echoes with the commentary in the proviso (a) above.

Comparison with the ICE Conditions

This ICE Clause 22 is very similar to the HKSAR Clause 22 except that the former has been more neatly arranged. One difference is that the ICE Clause expressly includes indemnity to injury or damage arising from "remedying any defects" while the HKSAR Clause may rely on the definition of the term "Works" to cover damage or injury arising from any remedial works to be carried out during the Maintenance Period. Another difference is that the ICE Clause does not separately provide for remedy in case the property of the Employer is damaged or any employee of the Employer is injured. The ICE Clause also allows for apportionment of liability in case the act or neglect of the Employer partly attributes to the damage. Additionally, when the exceptions apply such that the responsibility lies with the Employer, the ICE Clause allows for apportionment of damage in case the act or neglect of the Contractor partly contributes to the damage.

Comparison with the FIDIC Conditions

The FIDIC Clause 17.1(1) covers additionally disease and sickness which may not fall within the definition of "injury." It also expressly provides for mishaps arising from "remedying any defects." One major difference is that the FIDIC Clause excludes liability arising from the Employer's risks as defined in the FIDIC Sub-Clause 17.3, which covers similar items of "excepted risks" in the HKSAR Conditions. The liability is subject to apportionment if the other party also contributes to the loss.

 Q 17 These ground-level street lighting posts on the flyover construction site are erected by the lighting term contractor. What is the contractual position if one of the lighting posts falls and causes damage to a private car and to the adjacent columns under construction?

Clause 23

Design Responsibility

Except as may be provided for in the Contract, the Contractor shall not be responsible for the design of the permanent work or for the design of any Temporary Works designed by the Engineer.

Corresponding clauses in the ICE Conditions:

Clause 8

(2) The Contractor shall not be responsible for the design or specification of the Permanent Works or any part thereof (except as may be expressly provided in the Contract) or of any Temporary Works design supplied by the Engineer. The Contractor shall exercise all reasonable skill care and diligence in designing any part of the Permanent Works for which he is responsible.

Corresponding clauses in the FIDIC Conditions:

Clause 4.1, paragraph 3

... Except to the extent specified in the Contract, the Contractor (i) shall be responsible for all Contractor's Documents, Temporary Works, and such design of each item of Plant and Materials as is required for the item to be in accordance with the Contract, and (ii) shall not otherwise be responsible for the design or specification of the Permanent Works.

Clause 4.1 (c)

The Contractor shall be responsible for this part and it shall, when the Works are completed, be fit for the purposes for which the part is intended as are specified in the Contract.

條款 23

香港特別行政區土木工程合同一般條件：

設計的責任

除「合同」另有規定外，「承建商」不須負責設計「永久工程」或由「工程師」設計的任何「臨時工程」。

Commentary

For an ordinary civil engineering contract, the Contractor is only responsible for the design of Temporary Works while the permanent works is designed by the Employer's designer. However, it is not unusual that part of the permanent works design is specified to be carried out by the Contractor. These include relatively minor designs such as sub-frames and panels for noise barrier or water-points and piping system for landscape works. Sometimes, such design requirements are given in the particular specifications but it is recommended that if the design work is substantial, the requirements and responsibility should be detailed[1] in a Special Condition. If there is any requirement for the Contractor to coordinate the designs by sub-contractors, such as electrical and mechanical works, an express clause giving full details of co-ordinating activities must be included.[2]

Analysis and application

Clause 23 "… for the design of any Temporary Works designed by the Engineer."

If the Engineer wishes to design any Temporary Works and include those designs into the Contract, he must be extremely careful because any changes due to different Site conditions may amount to a variation. This is especially the case when the Engineer attempts to illustrate on the drawings how the permanent works are to be constructed. Such attempts may be construed by the Contractor as specifying the method of construction. Another area which needs special attention is temporary traffic diversion schemes. If such schemes are given on the Drawings, any changes to the schemes due to changes in traffic condition or different requirements of the authorities may amount to a variation.

Comparison with the ICE Conditions

The ICE Clause is similar to the HKSAR Clause. One difference is that it expressly requires that the Contractor shall exercise all reasonable skill, care and diligence in designing any part of the Permanent Works for which he is responsible. As stated in the commentary above, substantial Permanent Works design shall be provided by the Special Conditions in a HKSAR contract. The Special Conditions also require the Contractor to exercise all reasonable skill, care and diligence. The obligations in temporary works is, however, not limited to "reasonable skill, care and diligence" and the Contractor shall "take full responsibility for the adequate stability and safety of all operations on the Site" under the HKSAR Conditions Clause 20 or under the ICE Conditions Clause 8(3). See Max Abrahamson's commentary to the ICE Conditions in this respect.[3] Another difference is that the ICE Clause also explicitly refers to the responsibility for "Specification." The "design" in the HKSAR Conditions should generally cover also "Specification."

Comparison with the FIDIC Conditions

The FIDIC Clause describes what the responsibilities of the Contractor are and what responsibilities are not regarding the design and documentations of the Permanent and Temporary Works. More precise details are described such as "Contractor's Documents, Temporary Works, and such design of each item of Plant and Materials" probably because it is for international use. Clause 4.1 (c) is specifically included here for comparison because the design has to be fit for the purposes for which it is intended and is more stringent than the HKSAR and ICE requirements of "reasonable skill and care."

Notes

1. See Roger Knowles, *150 Contractual Problems and Their Solutions*, Section 1.9.
2. See Roger Knowles, *150 Contractual Problems and Their Solutions*, Section 1.3.
3. Max Abrahamson, *Engineering Law and the ICE Contracts*, 4th Edition.

Clause 25

Remedy on Failure to Insure

If the Contractor shall fail to effect and keep in force any insurance which he may be required to effect by any Special Condition of Contract then and in any such case the Employer may effect and keep in force any such insurance and pay such premiums as may be necessary for that purpose and such premiums together with expenses incurred shall be recoverable by the Employer from the Contractor.

Corresponding clauses in the ICE Conditions:

Clause 25

(3) If the Contractor shall fail upon request to produce to the Employer satisfactory evidence that there is in force any of the insurances required under the Contract then the Employer may effect and keep in force any such insurance and pay such premium or premiums as may be necessary for that purpose and from time to time deduct the amount so paid from any monies due or which may become due to the Contractor or recover the same as a debt due from the Contractor.

Corresponding clauses in the FIDIC Conditions:

Clause 18.1, paragraph 10

If the insuring Party fails to effect and keep in force any of the insurances it is required to effect and maintain under the Contract, or fails to provide satisfactory evidence and copies of policies in accordance with this Sub-Clause, the other Party may (at its option and without prejudice to any other right or remedy) effect insurance for the relevant coverage and pay the premiums due. The insuring Party shall pay the amount of these premiums to the other Party, and the Contract Price shall be adjusted accordingly.

條款 25

對不投購保險的補救

如「承建商」未能依據「合同」的「特別條件」投購所需要的任何保險及未能使其持續有效，「僱主」在此或在任何此類情況下可投購所需的保險及使其持續有效，及可為此支付所需的保費。「僱主」可向「承建商」追討「僱主」所付出的保費及由此引致的開支。

Commentary

Most contractors carrying out construction works are exposed to some types of risk such as weather condition or ground instability, and those may bring about damage to the Works or property and injury to the people working there or the third party such as the public. The contractors usually take out their own insurance including construction-all-risks insurance, which consists of insurance of the construction works and the third party injury or damage. As far as the Government of the HKSAR is concerned, most contracts require the Contractor to take out third party insurance, and until more recently the minimum amount to be insured is 10% of the Contract Sum. It is considered that the amount is purely arbitrary and bears no relation to the risk exposure. For more important or major construction contracts, the contract usually also specifies works insurance to cover the value of the works to be constructed. Now the *ETWB Technical Circular* requires an insurance advisor to advise on the risks and the amount to be insured as well as the policy wording. The insurance can either be controlled by the Contractor himself or by the Employer. The latter arrangement needs a long lead-time of around at least 18 months before construction commences and is suitable for multi-contracts and multi-disciplinary projects such as the MTRC and Airport Core Projects. For ordinary major projects, the contractor-controlled insurance is a simpler and more suitable arrangement. Apart from the construction-all-risks insurance, it is a legal requirement for the Contractor to take out the employees' compensation insurance for the workmen. Sometimes, the employees' compensation insurance may include the self-employed depending on the Contract requirements. Another type of insurance is the professional indemnity insurance, which is necessary when the Contractor or his designer carries out some design under the Contract. All such insurance coverage, if required by the Employer, would usually be spelt out in the Special Conditions of Contract. The Clause 25 simply provides for the unlikely situation when the Contractor fails to take out the required insurance under the Contract. Then the Employer can buy the insurance himself and recover the premium and all associated costs from the Contractor. In most cases, the Employer is reluctant to invoke this Clause because the lack of insurance does not have a direct impact on the Employer's finance. The insurance only caters for the situation when the loss or damage is so large that the Contractor is unable to bear the consequence, resulting in adverse effect on the progress of the Works or adverse financial consequences on the Employer. It is also quite often that if the Contractor is unable to place the required insurance, the Employer himself is unlikely to be able to do so.

Analysis and application

The interpretation of this Clause 25 is straightforward.

Comparison with the ICE Conditions

It should be noted that the ICE Conditions require the Contractor to take out Works and third party insurance and no Special Condition is needed. The ICE Clause 25(3) is similar to the HKSAR Clause 25 empowering the Employer to effect the insurance and recover the amount from the Contractor in case of the latter's default.

Comparison with the FIDIC Conditions

This FIDIC Clause 18.1 is very similar to the ICE Clause 25(3).

Clause 26

Accident or Injury to Workers

The Employer shall not be liable for or in respect of any damages or compensation payable at law in respect of or in consequence of any accident or injury to any worker or other person in the employ of the Contractor or any sub-contractor save and except an accident or injury resulting from any act or default of the Employer, his agents or employees and the Contractor shall indemnify and keep indemnified the Employer against all such damages and compensation, save and except as aforesaid and against all claims, demands, proceedings, damages, costs, charges and expenses whatsoever in respect thereof or in relation thereto.

Corresponding clauses in the ICE Conditions:

Clause 24

The Employer shall not be liable for or in respect of any damages or compensation payable at law in respect or in consequence of any accident or injury to any operative or other persons in the employment of the Contractor or any of his sub-contractors save and except to the extent that such accident or injury results from or is contributed to by any act or default of the Employer his agents or servants and the Contractor shall indemnify and keep indemnified the Employer against all such damages and compensation (save and except as aforesaid) and against all claims demands proceedings costs charges and expenses whatsoever in respect thereof or in relation thereto.

Corresponding clauses in the FIDIC Conditions:

Clause 18.4

The Contractor shall effect and maintain insurance against liability for claims, damages, losses and expenses (including legal fees and expenses) arising from injury, sickness, disease or death of any person employed by the Contractor or any other of the Contractor's Personnel.

The Employer and the Engineer shall also be indemnified under the policy of insurance, except that this insurance may exclude losses and claims to the extent that they arise from any act or neglect of the Employer or of the Employer's Personnel.

The insurance shall be maintained in full force and effect during the whole time that these personnel are assisting in the execution of the Works. For a Subcontractor's employees, the insurance may be effected by the Subcontractor, but the Contractor shall be responsible for compliance with this Clause.

條款 26

工人的意外或損傷

除非意外或損傷是由「僱主」、其代理人或僱員的行為或失責引致，對於任何受僱於「承建商」或任何「次承判商」的工人或其他人的任何意外或損傷而為此或因此構成依法律可得到的任何賠償或補償，「僱主」無須負法律責任。「承建商」亦必須保障「僱主」免除負責上述所有賠償及補償及一切因此而牽涉或與此有關的所有索償、要求、訴訟、賠償、訟費、收費及開支，並使「僱主」繼續可得到此保障。

Commentary

This Clause places upon the Contractor the liability for any damages or compensation arising from any accident or injury to workers or other employees of the Contractor or the sub-contractors except for accident or injury resulting from any act or default of the Employer, his agents or employees. The Contractor is required to indemnify the Employer against such loss. In Hong Kong, the Contractor has to take out employees' compensation insurance under the *Employees' Compensation Ordinance*.

Analysis and applications

Clause 26 "The Employer shall not be liable for or in respect of any damages or compensation payable at law …"

An injured worker can sue for compensation under the *Employees' Compensation Ordinance* or under the common law. Under the Ordinance, there is a statutory limit but there is no limit under the common law. The Employer shall not be liable for any sum awarded on the above basis. However, if there is any settlement sum paid by the Employer, the Contractor shall only be liable to the extent payable at law.

Clause 26 "… except an accident or injury resulting from any act or default of the Employer, his agents or employees …"

Unlike construction-all-risks insurance for the Works and third party damage, which is taken out in the joint name of the Employer and the Contractor, the employees' compensation insurance is compulsory under the law and is taken out in the name of the Contractor only. The Engineer, as the agent of the Employer, may be liable if a workman is injured arising from his default, e.g., in giving inappropriate instructions and therefore he has to take out his own professional indemnity insurance.

Comparison with the ICE Conditions

This ICE Clause 24 is very similar to the HKSAR Clause 26 except that the former expressly allows for apportionment of damages or compensation in case the accident is partly arising from Employer's default. Under the HKSAR Clause, apportionment may be made under the law.

Comparison with the FIDIC Conditions

This FIDIC Clause 18.4 is also similar to the HKSAR Clause. The former expressly covers sickness and disease, as well as legal fees and expenses. It also obliges the Contractor to take out insurance for his employees covering also any legal fees and expenses while in the HKSAR, employees' insurance is compulsory under the law.

The HKSAR Conditions:

Compliance with Enactments and Regulations

The Contractor shall conform in all respects with:

(a) the provisions of any enactment,

(b) the regulations or bye-laws of any local or other duly constituted authority, and

(c) the rules and regulations of such public bodies and statutory authorities as are referred to in Clause 29,

and any additions or amendments thereto during the continuance of the Works, which are applicable to the Works, and shall keep the Employer indemnified against all penalties and liabilities of every kind for breach of any such enactment, regulations, bye-laws or rules.

Corresponding clauses in the ICE Conditions:

Clause 26

(3) The Contractor shall ascertain and conform in all respects with the provisions of any general or local Act of Parliament and the Regulations and Bye-laws of any local or other statutory authority which may be applicable to the Works and with such rules and regulations of public bodies and companies as aforesaid and keep the Employer indemnified against all penalties and liability of every kind for breach of any such Act Regulation or Bye-law. Provided always that

(a) the Contractor shall not be required to indemnify the Employer against the consequences of any such breach which is the unavoidable result of complying with the Contract or instructions of the Engineer

(b) if the Contract or instructions of the Engineer shall at any time be found not to be in conformity with any such Act Regulation or Bye-law the Engineer shall issue such instructions including the ordering of a variation under Clauses 51 as may be necessary to ensure conformity with such Act Regulation or Bye-law and

(c) the Contractor shall not be responsible for obtaining any planning permission which may be necessary in respect of the Permanent Works in their final position or of any Temporary Works designed by the Engineer in their designated position on Site. The Employer hereby warrants that all such permissions have been or will in due time be obtained.

Corresponding clauses in the FIDIC Conditions:

Clause 1.13

The Contractor shall, in performing the Contract, comply with applicable Laws. Unless otherwise stated in the Particular Conditions:

(a) the Employer shall have obtained (or shall obtain) the planning, zoning or similar permission for the Permanent Works, and any other permissions described in the Specification as having been (or being) obtained by the Employer; and the Employer shall indemnify and hold the Contractor harmless against and from the consequences of any failure to do so; and

(b) the Contractor shall give all notices, pay all taxes, duties and fees, and obtain all permits, licences and approvals, as required by the Laws in relation to the execution and completion of the Works and the remedying of any defects; and the Contractor shall indemnify and hold the Employer harmless against and from the consequences of any failure to do so.

Clause 13.7

The Contract Price shall be adjusted to take account of any increase or decrease in Cost resulting from a change in the Laws of the Country (including the introduction of new Laws and the repeal or modification of existing Laws) or in the judicial or official governmental interpretation of such Laws, made after the Base Date, which affect the Contractor in the performance of obligations under the Contract.

If the Contractor suffers (or will suffer) delay and/or incurs (or will incur) additional Cost as a result of these changes in the Laws or in such interpretations, made after the Base Date, the Contractor shall give notice to the Engineer and shall be entitled subject to Sub-Clause 20.1 [*Contractor's Claims*] to:

(a) an extension of time for any such delay, if completion is or will be delayed, under Sub-Clause 8.4 [*Extension of Time for Completion*], and

(b) payment of any such Cost, which shall be included in the Contract Price.

After receiving this notice, the Engineer shall proceed in accordance with Sub-Clause 3.5 [*Determinations*] to agree or determine these matters.

條款 30

香港特別行政區土木工程合同一般條件：

遵從法例條文及規例

在「本工程」繼續期間，「承建商」須在各方面遵從以下適用於「本工程」的：

(a) 任何法例條文；

(b) 任何地方當局或其他依法組成的機構的規例或附例；及

(c) 在第29條所提及的公共機構及法定機構的規則及規例，

以及對以上各項所作出的任何增補或修改。「承建商」也必須保障「僱主」免除為所有因違反任何此等法例、規例、附例或規則而引致的所有各種罰則及法律責任，並使「僱主」繼續可得到此保障。

Commentary

This Clause expressly requires the Contractor to comply with the law and any addition and amendment to the law during the continuance of the Works. One well-known example was the passing of the legislation in about 1980 prohibiting the use of power mechanical plant during Sunday and other Public Holidays under which many contractors of the on-going contracts suffered delay to their works and incurred additional expenses. In principle, the Contractor cannot recover any Cost nor is he entitled to any extension of time under this Clause. In line with international practice, the Government of the HKSAR now adopts a recommendation of the *Jesse Grove Report* to allow the Contractor to claim any extra time or additional expenses incurred as a result of any change in law by way of a Special Condition.[1] Change in law is defined in the said Special Condition as an addition or amendment of any enactment, regulation, bye-law or rules listed in the said Special Condition made on or after 10 days before the Tender closing date.

Analysis and application

Clause 30(b) "the regulation or bye-laws of any local or duly constituted authority, ..."

Regulation is the same as subsidiary legislation, which is usually technical in nature drafted by a department through delegated authority, for example, the Roads Traffic (Traffic Control) Regulations. Bye-laws of any local or duly constituted authorities could mean bye-laws or regulations of any non-government authority, for example, the Construction Industry Training Authority.

Clause 30(c) "the rules and regulations of such public bodies and statutory authorities ..."

Public bodies include the Executive Council and Legislative Council, municipal council and any department of the Government. Statutory authorities could mean Government departments, such as the Waterworks Authority, Mines Division or Transport Department, or the non-government Hong Kong Airport Authority, which are carrying out statutory functions.

Clause 30 "... and any additions or amendments thereto during the continuance of the Works, ..."

This means the Contractor shall conform with any change in law and the Clause does not allow for any reimbursement of Costs incurred as a result of such changes. However, as discussed in the commentary, a Special Condition is now adopted to allow for extension of time and reimbursement of expenses due to changes in law.

Clause 30 "… and shall keep the Employer indemnified against all penalties and liabilities of every kind for breach of any such enactment, regulation, bye-laws or rules."

If the Contractor performs any act or omit to duly perform any act resulting in the Employer being sued for any damages or being imposed any penalties by the authorities, e.g., breach of any safety or environmental legislations, the Contractor shall be responsible for any costs incurred by the Employer.

Comparison with the ICE Conditions

This ICE Clause 26(3) also places the responsibility of complying with the law on the Contractor but there are the following differences. Firstly, this ICE Clause does not say anything about changes in legislation after the tender is closed. Secondly, it lists three conditions under which the Cost of or the consequences of any breach will be shifted to the Employer, namely, (a) the unavoidable result of complying with the Contract or the instruction of the Engineer, (b) the need to rectify any non-compliance with the legislation resulting in a variation order and, (c) obtaining any planning permission for Permanent Works or Temporary Works designed by the Engineer. This compensates to a certain extent of any changes in legislation rendering part of the Contract or the Engineer's instruction unlawful, which could only be rectified with addition time and Cost.

Comparison with the FIDIC Conditions

Like the ICE Clause, the FIDIC Clause 1.13 requires the Contractor to comply with the Law but requires the Employer to obtain any planning permission for the Permanent Works or any other permission described in the Specification. The FIDIC Clause 13.7 allows for price adjustment and extension of time to make up for delay arising from a change in the Laws of the Country.

Notes

1.   See *Recommendations in Grove Report on General Conditions of Contract.*

 This is an elevated platform of a tower construction site. A legislation is passed that moving falsework of area exceeding 50m² must be personally inspected and signed by two qualified structural engineers. What is the Contractor's position on the extra expenses?

 What are the legislations or bye-laws which are applicable to this major construction site located in the city centre?

Clause 46

Removal of Unsatisfactory Material and Work

(1) The Engineer shall during the progress of the Works have the power to order in writing:
   (a) the removal from the Site within such time as may be specified in the order of any material which in the opinion of the Engineer is not in accordance with the Contract,
   (b) the substitution of proper and suitable material, and
   (c) the removal and proper re-execution, notwithstanding any previous examination, measurement or test thereof or any interim payment therefor, of any work which, in respect of materials or workmanship, is not in accordance with the Contract.

(2) The Contractor shall bear the expense of uncovering, breaking up and removal from the Site of any material or work not in accordance with the Contract and the Contractor shall also bear the expense of reinstating and making good all consequential damage to the Works resulting from such uncovering, breaking up or removal.

(3) Where the rectification of any work or replacement of any material by the Contractor which does not comply with the Contract would involve the removal and re-execution of the original permanent work the Engineer may but shall not be obliged to give directions for a variation of the Works in lieu of such removal and re-execution at no additional expense to the Employer.

   Provided that if in the opinion of the Engineer such variation has involved the Contractor in expense in excess of that which would have been involved in the removal and re-execution of the original permanent work then the Engineer shall value such excess in accordance with Clause 61, and shall certify in accordance with Clause 79.

(4) In the event that the Engineer exercises any of his powers under sub-clause (1) of this Clause concerning materials supplied by the Employer, and if in the opinion of the Engineer the Contractor could not have reasonably ascertained that the material was not in accordance with the Contract then the Engineer shall ascertain the Cost incurred, and shall certify in accordance with Clause 79.

Corresponding clauses in the ICE Conditions:

Clause 39

(1) The Engineer shall during the progress of the Works have power to instruct in writing the
   (a) removal from the Site within such time or times specified in the instruction of any materials which in the opinion of the Engineer are not in accordance with the Contract
   (b) substitution with materials in accordance with the Contract and
   (c) removal and proper replacement (notwithstanding any previous test thereof or interim payment therefor) of any work which in respect of
      (i) material or workmanship or
      (ii) design by the Contractor or for which he is responsible

is not in the opinion of the Engineer in accordance with the Contract.

(2) In case of default on the part of the Contractor in carrying out such instruction the Employer shall be entitled to employ and pay other persons to carry out the same and all costs consequent thereon or incidental thereto as determined by the Engineer shall be recoverable from the Contractor by the Employer and may be deducted by the Employer from any monies due or to become due to him and the Engineer shall notify the Contractor accordingly with a copy to the Employer.

(3) Failure of the Engineer or any other person acting under him pursuant to Clause 2 to disapprove any work or materials shall not prejudice the power of the Engineer or any such person subsequently to take action under this Clause.

Corresponding clauses in the FIDIC Conditions:

Clause 7.6

Notwithstanding any previous test or certification, the Engineer may instruct the Contractor to:

(a) remove from the Site and replace any Plant or Materials which is not in accordance with the Contract,

(b) remove and re-execute any other work which is not in accordance with the Contract, and

(c) execute any work which is urgently required for the safety of the Works, whether because of an accident, unforeseeable event or otherwise.

The Contractor shall comply with the instruction within a reasonable time, which shall be the time (if any) specified in the instruction, or immediately if urgency is specified under sub-paragraph (c).

If the Contractor fails to comply with the instruction, the Employer shall be entitled to employ and pay other persons to carry out the work. Except to the extent that the Contractor would have been entitled to payment for the work, the Contractor shall subject to Sub-Clause 2.5 [*Employ's Claims*] pay to the Employer all costs arising from this failure.

條款 46

香港特別行政區土木工程合同一般條件：

移走不合意的物料及工程

(1) 在「本工程」施行期間，「工程師」有權以書面命令：

    (a) 按指定的時間內將「工程師」認為不符合「合同」規定的任何物料搬出「工地」；

    (b) 用恰當及適合的物料代替；及

    (c) 移去任何因物料或技術水平不符合「合同」規定的工程，並重新妥善地實施此工程，雖然以前已進行了任何檢查、計量或測試，或支付中期付款。

(2) 「承建商」必須負擔回收、拆除及移走「工地」內任何不符合「合同」規定的物料或工程的開支，亦必須負擔回復原狀及修復因上述回收、拆除或移走行動而引致對「本工程」構成的一切損害的開支。

(3) 如由於「承建商」未能按照「合同」規定施工而因此須糾正任何工程或更換任何物料致原來的「永久工程」要被移走及重新實施，「工程師」可（但無責任）發出「本工程」變更的指示，以取代移走及重新實施此等工程而無須令「僱主」負擔任何額外開支。

    但如「工程師」認為此項工程變更已令「承建商」牽涉的開支超越移走及重新實施原先的「永久工程」的開支，「工程師」則必須根據第61條對超越的開支釐定價值，並根據第79條予以核實。

(4) 如「工程師」行使他在本條第(1)款下的任何權力於「僱主」所供應的物料，而且他認為「承建商」是不可能合理地確定這此物料是不符合「合同」的規定，則「工程師」必須核實所引致的成本，並根據第79條予以核實。

Commentary

This Clause empowers the Engineer to order the removal of any unsatisfactory materials or any unsatisfactory work and the substitution with satisfactory materials or proper re-execution of the work respectively. For the removal of unsatisfactory materials, the Engineer can specify the time for removal such that the materials may not be used intentionally or mistakenly in the Works. For the removal of the unsatisfactory work, any previous examination and test or interim payment will not prejudice the Engineer from making a removal and re-execution order. The Contractor has to bear all the financial and time consequences. However, it is quite common that the removal of permanent work may cause unacceptable delay to the completion and in such case the Engineer may direct a variation of the Works instead of complete re-execution, e.g., for a bored pile below the specified capacity, an additional smaller pile can be placed and the pile cap can be modified. The Government now uses a Special Condition, which empowers the Engineer to accept substandard Works (which would not affect the future functioning) and determine a deduction of the Contract price.

Analysis and application

Clause 46(1) "The Engineer shall during the progress of the Works ..."

The wording "during the progress of the Works" may arguably suggest that the Clause applies to the construction period only. For rectification of defective works during the Maintenance Period, the Engineer can invoke Clause 56(2) and (3) on execution of work of repair.

Clause 46(1)(c) "... notwithstanding any previous examination, measurement or test thereof or any interim payment therefor, ..."

In general, the Contractor's obligation to comply with the Contract provisions will not be affected by the Engineer's or Engineer's Representative's previous act or omission in accordance with Clause 2(4) or the previous approval of documents submitted by the Contractor under Clause 7(5). There may be an argument that the previous approval of substandard works or materials constitutes a variation to works or materials to a lower standard. Unless the Contractor confirms in writing subsequently that a variation has been ordered and that order was made before the variation takes place, such approval by the Engineer may not bind the Employer.

Clause 46(3) "... would involve the removal and re-execution of the original permanent works the Engineer may but shall not be obliged to give directions for a variation of the Works in lieu of such removal and re-execution at no additional expense to the Employer."

The Engineer may have an option to give direction for variation of the Works in lieu of removal to ameliorate the grave financial and time consequences. This should be wide enough to cover cases where the Contractor submits remedial proposals which amounts to a variation of the Works and the Engineer gives approval. The Engineer normally requires the proposals to be checked by an independent checking engineer. For major remedial works involving substantial re-design by the Contractor, the future responsibility of the re-design should be clarified and this should preferably also involve an agreement between the Employer and the Contractor.

Clause 46(3) "... such variation has involved the Contractor in expense in excess of that which would have been involved in the removal and re-execution of the original permanent works then the Engineer shall value such excess ..."

If a variation to the Works affected by defective workmanship, etc. is ordered, this proviso defines the financial limit which the Contractor is liable, i.e., the Cost of removal and re-execution of the original permanent works but there is no provision for any saving as a result of the direction of the Engineer for a variation. It should be noted that the variation should presumably not result in a lower standard of the overall work. As explained in the commentary, a Special Condition adopted by the Government allows acceptance of substandard work which would not compromise its future performance and the Engineer can determine a deduction of the Contract price.

Clause 46(4) "... materials supplied by the Employer, ... the Contractor could not have reasonably ascertained that the material was not in accordance with the Contract then the Engineer shall ascertain the Cost incurred ..."

For materials supplied by the Employer for incorporation into the Works, it is very important that all tests are completed satisfactorily and the materials are thoroughly inspected by the Contractor in the presence of the Engineer's Representative. Where appropriate, the Contractor shall be obliged to sign a document stating that all materials are to the Contractor's satisfaction. However, for any latent defects, this sub-clause may still be invoked.

Comparison with the ICE Conditions

The ICE sub-clause 39(1) is very similar to the HKSAR sub-clause 46(1) except that the former also covers any "design by the Contractor or for which he is responsible." The HKSAR Conditions now relies on a Special Condition concerning design by the Contractor under which the Contractor is responsible for making good any defects arising from his own design. The ICE sub-clause 39(2) allows the Employer to carry out any replacement work in case the Contractor refuses or fails to comply with the instruction of the Engineer while the HKSAR Conditions relies on Clause 82, which covers both non-urgent and urgent cases.

Comparison with the FIDIC Conditions

This FIDIC Clause 7.6 is more logically arranged and its application is not limited to "during the progress of the Works" and would therefore arguably cover both the Construction and Defects Notification Period. It should be noted, however, that the FIDIC Conditions has a long and substantial Clause (sub-clause 11.1 to 11.6) covering remedial works during the Defects Notification Period. The FIDIC Clause 7.6 covers unsatisfactory materials and plant, which are part of the Permanent Works and any other completed Works that is not in accordance with the Contract. Additionally, the Contractor is obliged to execute any work which is urgently required for the safety of the Works. For all the above, the Contractor has to comply with the instruction within a reasonable time or immediately if urgency is specified, unlike the HKSAR Conditions and the ICE Conditions, which only requires unsatisfactory materials to be removed within a reasonable time. The FIDIC Clause also allows the Employer to arrange his own persons to carry out the repair works if the Contractor fails to do so and recover Cost from the Contractor as the Employer's Claims.

 The contractor has commenced the construction of this gantry before the steel structural strength is tested. The structural strength is subsequently found to be slightly below the requirements. What courses of action are opened to the Engineer?

Clause 47

Commencement of the Works

The Contractor shall commence the Works on the date for commencement of the Works as notified in writing by the Engineer and shall proceed with the same with due diligence. The date so notified by the Engineer shall be within the period of time after the date of acceptance of the Tender as stated in the Appendix to the Form of Tender. The Contractor shall not commence the Works before the notified date for commencement.

Corresponding clauses in the ICE Conditions:

Clause 41

(1) The Works Commencement Date shall be

(a) the date specified in the Appendix to the Form of Tender or if no date is specified

(b) a date between 14 and 28 days of the award of the Contract to be notified to the Contractor by the Engineer in writing or

(c) such other date as may be agreed between the parties.

(2) The Contractor shall start the Works on or as soon as is reasonably practicable after the Works Commencement Date. Thereafter the Contractor shall proceed with the Works with due expedition and without delay in accordance with the Contract.

Corresponding clauses in the FIDIC Conditions:

Clause 8.1

The Engineer shall give the Contractor not less than 7 days' notice of the Commencement Date. Unless otherwise stated in the Particular Conditions, the Commencement Date shall be within 42 days after the Contractor receives the Letter of Acceptance.

The Contractor shall commence the execution of the Works as soon as is reasonably practicable after the Commencement Date, and shall then proceed with Works with due expedition and without delay.

條款 47

「本工程」的開始

「承建商」必須依照「工程師」以書面通知的「本工程」開工日期開始「本工程」，及必須盡力施行「本工程」。「工程師」通知的日期必須是在「投標表格」的「附件」中所載述的「投標」獲接納日期後的一段時間內。「承建商」不可在獲通知的開工日期前開始「本工程」。

Commentary

For the HKSAR civil engineering contracts, the Contractor is usually given 14 days after the letter of acceptance to commence the Works and the times for completion then run from that date. For a very tight programme, the period of commencement may be shortened to 7 days. To commence the Works within 14 days is more reasonable because the Contractor has to arrange his insurance and mobilize his resources before commencing any works on the Site. During this period, the Contractor must very quickly produce a programme for the Works under Clause 16 of the HKSAR Conditions or an initial Works Programme (under a Special Condition for Mega Project Contracts).

Analysis and application

Clause 47 "… on the date for commencement of the Works as notified in writing by the Engineer …"

Notification may be given in the letter of acceptance if the tender is accepted by the Consultant on behalf of the Employer or in a separate letter by the Engineer after acceptance of Tender by the Employer.

Clause 47 "… and shall proceed with the same with due diligence …"

It is suggested that commencement must be real and substantial which may take place on Site or off-Site depending on the circumstances. For example, if the Site is along a stretch of busy road, or in a densely populated area, the arrangement for traffic diversion or baseline monitoring of environmental data may take at least several weeks before substantial Works can commence on the Site.

Clause 47 "… The date so notified by the Engineer shall be within the period of time after the date of acceptance of the Tender as stated in the Appendix to the Form of Tender …"

Though the Engineer can give a notice period less than the period of time stated in the Appendix to the Form of Tender, he should be mindful of the time required for taking out insurance and the Contractor's mobilization.

Comparison with the ICE Conditions

The ICE Clause 41 is similar to the HKSAR Clause 47 but it specifies a date of commencement between 14 days and 28 days of the award of Contract (which means the same as the issue of the letter of acceptance) or any other date as may be agreed. This is considered to be a more practical and equitable arrangement.

Comparison with the FIDIC Conditions

The FIDIC sub-clause is also similar but it specifies a date of commencement between 7 days and 42 days after the letter of acceptance, to cater for different situations in different countries.

Clause 48

Possession of the Site

(1) Save in so far as the Contract may prescribe the extent of Portions of the Site of which the Contractor is to be given possession from time to time and the order in which such Portions shall be made available to him and, subject to any requirement in the Contract as to the order in which the Works shall be executed, the Employer shall give to the Contractor on the date for commencement notified by the Engineer in accordance with Clause 47 possession of so much of the Site as may be required to enable the Contractor to commence and proceed with the construction of the Works in accordance with the programme referred to in Clause 16 and otherwise in accordance with such reasonable proposals in writing as the Contractor shall make to the Engineer. The Employer will from time to time, as the Works proceed, give to the Contractor possession of such further parts of the Site as may be required to enable the Contractor to proceed with construction of the Works with due despatch in accordance with the said programme or proposals, as the case may be.

(2) If upon written application having been made by the Contractor to the Engineer, the Engineer is of the opinion that the Contractor has been involved in additional expenditure by reason of the progress of the Works or any part thereof having been materially affected by the failure of the Employer to give possession in accordance with this Clause then the Engineer shall ascertain the Cost incurred, and shall certify in accordance with Clause 79.

(3) The Contractor shall bear all expenses and charges for special or temporary wayleaves required by him in connection with access to the Site.

Corresponding clauses in the ICE Conditions:

Clause 42

(1) The Contract may prescribe

   (a) the extent of portions of the Site of which the Contractor is to be given possession from time to time

   (b) the order in which such portions of the Site shall be made available to the Contractor

   (c) the availability and the nature of the access which is to be provided by the Employer

   (d) the order in which the Works shall be constructed.

(2) (a) Subject to sub-clause (1) of this Clause the Employer shall give to the Contractor on the Works Commencement Date possession of the whole of the Site together with such access thereto as may be necessary to enable the Contractor to commence and proceed with the construction of the Works.

   (b) Thereafter the Employer shall during the course of the Works give to the Contractor such further access in accordance with the Contract as is necessary to enable the Contractor to proceed with the construction of the Works with due dispatch.

(3) If the Contractor suffers delay and/or incurs extra cost from failure on the part of the Employer to give possession or access in accordance with the terms of this Clause the Engineer shall take such delay into account in determining any extension of time to which the Contractor is entitled under Clause 44 and the Contractor shall subject to Clause 53 be paid in accordance with Clause 60 the amount of extra Cost to which he may be entitled. Profit shall be added thereto in respect of any additional Permanent or Temporary Works.

Corresponding clauses in the FIDIC Conditions:

Clause 2.1

The Employer shall give the Contractor right of access to, and possession of, all parts of the Site within the time (or times) stated in the Appendix to Tender. The right and possession may not be exclusive to the Contractor. If, under the Contract, the Employer is required to give (to the Contractor) possession of any foundation, structure, plant or means of access, the Employer shall do so in the time and manner stated in the Specification. However, the Employer may withhold any such right or possession until the Performance Security has been received.

If no such time is stated in the Appendix to Tender, the Employer shall give the Contractor right of access to, and possession of, the Site within such times as may be required to enable the Contractor to proceed in accordance with the programme submitted under Sub-Clause 8.3 [*Programme*].

If the Contractor suffers delay and/or incurs Cost as a result of a failure by the Employer to give any such right or possession within such time, the Contractor shall give notice to the Engineer and shall be entitled subject to Sub-Clause 20.1 [*Contractor's Claims*] to:

(a) an extension of time for any such delay, if completion is or will be delayed, under Sub-Clause 8.4 [*Extension of Time for Completion*], and

(b) payment of any such Cost plus reasonable profit, which shall be included in the Contract Price.

After receiving this notice, the Engineer shall proceed in accordance with Sub-Clause 3.5 [*Determinations*] to agree or determine these matters.

However, if and to the extent that the Employer's failure was caused by any error or delay by the Contractor, including an error in, or delay in submission of, any of the Contractor's Documents, the Contractor shall not be entitled to such extension of time, Cost or profit.

條款 48

香港特別行政區土木工程合同一般條件：

「工地」的管用

(1) 依「合同」規定要不時給予「承建商」「工地」分區管用權的範圍及次序，及在符合「合同」所規定「本工程」實施的次序下，「僱主」必須在「工程師」根據第47條通知的開工的日期將足夠「工地」的管用權給予「承建商」，使「承建商」能按照第16條所提及的施工計劃開展及施行「本工程」。「僱主」在「本工程」施行期間必須不時給予「承建商」「工地」的其他有需要的部分的管用權，以使「承建商」能按照上述施工計劃或根據「承建商」以書面給予「工程師」的合理提議（視乎情況而定）以應有的速度實施「本工程」。

(2) 如「承建商」經書面向「工程師」提交申請而「工程師」認為是由於「僱主」未能按照本條給予「承建商」管用權，以致「本工程」或其任何部份的進展實質受到影響而使「承建商」牽涉額外開支，則「工程師」必須確定因此引致的成本，並必須根據第79條予以核實。

(3) 「承建商」必須承擔為進入「工地」時所需要的特殊或臨時道路通行權的所有有關開支及收費。

Commentary

This Clause obligates the Employer to give possession of the various Portions of the Site to the Contractor for the carrying out of the Works in accordance with the times and requirements stipulated in the Contract. However, the Contractor may not have any valid claim for delay in possession unless he can show by his Clause 16 programme or his reasonable proposal previously submitted in writing that the progress of Works has been materially affected, involving the Contractor in extra time and cost. It must be noted that even after having possession of any Portion of the Site, the Contractor's carrying out of the Works is still subject to the requirements of the Contract and the various authorities. For example, if the Works are carried out along a public road, the Contractor has to submit traffic management proposals to the Hong Kong Police Force, Transport Department and Highways Department, etc. for approval and then obtain an excavation permit. If the Works involve reclamation, the operations are subject to the approval of the Marine Department and the *Dumping at Sea Ordinance*. Sometimes, the Works are to be carried out in an existing engineering works itself, such as a sewage or water treatment plant, and in such case, both the gaining of access and the possession of the working area are equally important. In other cases, there are no definite areas to possess, for example, the installation of the Traffic Control and Surveillance System on a trunk road. It is not unusual that both the access and the Site itself have to be shared between several contractors working at the same time. All the above arrangements and requirements need to be carefully written into the Contract.

Analysis and application

Clause 48(1) "… subject to any requirement in the Contract as to the order in which the Works shall be executed, …"

Sometimes the Contract may specify that certain part or Section of the Works is to be constructed first before carrying out the remaining parts of the Works. Before the completion of that certain part or Section, the Contractor may not need to be given possession of the remaining Portion of the Site, even though the Contract may have prescribed a possession date. In any case, the Contractor's programme should have catered for such construction sequences.

Clause 48(1) "… the Employer shall give to the Contractor … possession of so much of the Site as may be required to enable the Contractor to commence and proceed with the construction of the Works in accordance with the programme referred to in Clause 16 and otherwise in accordance with such reasonable proposals in writing …"

Even if the Contract has prescribed the possession dates of various Portions of the Site, the Employer is only obliged to give possession of the extent of the Site required for carrying out the Works in accordance with the Clause 16 programme. In other words, even the Employer is not able to give the whole Portion of the Site to the Contractor in the prescribed time, the Contractor may not be entitled to any extension of time nor any prolongation cost if the Works within the Portion are not yet scheduled to be carried out according to the programme. The wording "otherwise in accordance with such reasonable proposals in writing" means that even if the Contractor proposes to carry out the Works, for example, earlier than the time described in the programme, the Employer is obliged to give possession of the relevant Portion of the Site provided that the proposals are reasonable. The reasonable proposals may also include areas for fabricated steel components, precast concrete works or storage.

Clause 48(1) "… give to the Contractor possession of such further parts of the Site as may be required to enable the Contractor to proceed with construction of the Works …"

Any part or Portion of the Site not yet given to the Contractor in the first instance has to be given at a later time to suit the Contractor's construction operation.

Clause 48(2) "… the Engineer is of the opinion that the Contractor has been involved in additional expenditure by reason of the progress of the Works or any part thereof having been materially affected by the failure of the Employer to give possession …"

The Contractor shall only be entitled to Cost, which includes overhead but not profit, if the Contractor incurs additional expenditure because the progress of the Works has been materially affected by the failure of the Employer to give possession of the Site. In other words, if the progress of Works is not materially affected, e.g., the Contractor is still waiting for the arrival of the pipes he has ordered, he will not be entitled to any Cost even if the Employer fails to give possession of any part or Portion of the Site in the prescribed time. The sub-clause 50(1)(b)(vi) gives entitlement of extension of time to the Contractor for non-possession of the Site but not Cost and the claim for associated prolongation cost has to be made under this sub-clause.

Clause 48(3) "The Contractor shall bear all expenses and charges for special or temporary wayleaves required by him in connection with access to the Site."

Normal access to the Site is usually described in the Specification or Drawings and special or temporary wayleaves could mean some further access to the place of construction which the Contractor may like to acquire to facilitate his construction, e.g., by arrangement with private parties.

Comparison with the ICE Conditions

This ICE Clause 42 is essentially the same as the HKSAR Clause 48 except that the former is more neatly arranged. The ICE Clause in addition suggests writing into the Contract the availability and nature of access to be provided by the Employer which is not expressly mentioned in the HKSAR Clause. The ICE Clause also allows, as a result of non-possession, any extra costs incurred. If additional Permanent or Temporary Works are required as a result of non-possession of the Site, profit can be allowed for the additional Works.

Comparison with the FIDIC Conditions

Under this FIDIC Clause 2.1, the Employer shall give the Contractor both the right of access and possession of the Site, or part of the Site within the time stated in the Appendix to Form of Tender but qualifies that the right and possession may not be exclusive to the Contractor. Under the FIDIC Clause 4.15, the Contractor is deemed to have satisfied himself on the access conditions. The possession may also include foundation, structure and plant or means of access in which further Works may be carried out. A major difference of this FIDIC Clause from the HKSAR Clause and the ICE Clause is that possession may be withheld until the Performance Security has been received. For non-possession of the Site, the Contractor shall be entitled to Cost plus reasonable profit.

 Q 21 The Contract states that the overhead cables will be diverted when the Site is given to the Contractor. It is subsequently decided that such cables must be retained and the Contractor has to complete the viaduct works under such condition. What is the contractual position? What Clauses can the Contractor rely upon?

Clause 50

Extension of Time for Completion

(1) (a) As soon as practicable but in any event within 28 days after the cause of any delay to the progress of the Works or any Section thereof has arisen, the Contractor shall give notice in writing to the Engineer of the cause and probable extent of the delay.

    Provided that as soon as the Contractor can reasonably foresee that any order or instruction issued by the Engineer is likely to cause a delay to the progress of the Works or any Section thereof the Contractor shall forthwith give notice in writing to the Engineer and specify the probable effect and extent of such delay. Such notice shall not in any event be given later than 28 days after the Engineer has issued the relevant order or instruction.

  (b) If in the opinion of the Engineer the cause of the delay is

    (i) inclement weather and/or its consequences adversely affecting the progress of the Works,

    (ii) the hoisting of tropical cyclone warning signal No. 8 or above, or

    (iia) a Black Rainstorm Warning, or

    (iii) an instruction issued by the Engineer under Clause 5, or

    (iv) a variation ordered under Clause 60, or

    (v) a substantial increase in the quantity of any item of work included in the Contract not resulting from a variation ordered under Clause 60, or

    (vi) the Contractor not being given possession of the Site or any Portion or part thereof in accordance with the Contract or is subsequently deprived of it by the Employer, or

    (vii) a disturbance to the progress of the Works for which the Employer or the Engineer or a Specialist Contractor is responsible including but not restricted to any matter referred to in Clause 63, or

    (viii) the Engineer suspending the Works in accordance with Clause 54 in so far as the suspension is not occasioned by the circumstances described in Clause 54(2)(a) to (d), or

    (ix) any utility undertaking or other duly constituted authority failing to commence or to carry out in due time any work directly affecting the execution of the Works, provided that the Contractor has taken all practical steps to cause the utility undertaking or duly constituted authority to commence or to proceed with such work, or

    (x) delay on the part of any Nominated Sub-contractor for any reason specified in sub-clauses (b)(i) to (ix) of this Clause and which the Contractor has taken all reasonable steps to avoid or reduce, or

    (xi) any special circumstance of any kind whatsoever,

    then the Engineer shall within a reasonable time consider whether the Contractor is fairly entitled to an extension of time for the completion of the Works or any Section thereof.

Corresponding clauses in the ICE Conditions:

Clause 44

(1) Should the Contractor consider that

  (a) any variation ordered under Clause 51(1) or

  (b) increased quantities referred to in Clause 51(4) or

  (c) any cause of delay referred to in these Conditions or

  (d) exceptional adverse weather conditions or

  (e) any delay impediment prevention or default by the Employer or

  (f) other special circumstances of any kind whatsoever which may occur

be such as to entitle him to an extension of time for the substantial completion of the Works or any Section thereof he shall within 28 days after the cause of any delay has arisen or as soon thereafter as is reasonable deliver to the Engineer full and detailed particulars in justification of the period of extension claimed in order that the claim may be investigated at the time.

(2) (a) The Engineer shall upon receipt of such particulars consider all the circumstances known to him at that time and make an assessment of the delay (if any) that has been suffered by the Contractor as

a result of the alleged cause and shall so notify the Contractor in writing.

(b) The Engineer may in the absence of any claim make an assessment of the delay that he considers

has been suffered by the Contractor as a result of any of the circumstances listed in sub-clause (1) of this Clause and shall so notify the Contractor in writing.

Corresponding clauses in the FIDIC Conditions:

Clause 8.4

The Contractor shall be entitled subject to Sub-Clause 20.1 [*Contractor's Claims*] to an extension of the Time for Completion if and to the extent that completion for the purposes of Sub-Clause 10.1 [*Taking Over of the Works and Sections*] is or will be delayed by any of the following causes:

(a) a Variation (unless an adjustment to the Time for Completion has been agreed under Sub-Clause 13.3 [*Variation Procedure*]) or other substantial change in the quantity of an item of work included in the Contract,

(b) a cause of delay giving an entitlement to extension of time under a Sub-Clause of these Conditions,

(c) exceptionally adverse climatic conditions,

(d) Unforeseeable shortages in the availability of personnel or Goods caused by epidemic or governmental actions, or

(e) any delay, impediment or prevention caused by or attributable to the Employer, the Employer's Personnel, or the Employer's other contractors on the Site.

條款 50

香港特別行政區土木工程合同一般條件：

延長竣工限期

(1) (a) 倘若「本工程」或其任何「工段」的進展受到延誤，「承建商」必須在實際可行的情況下儘快，但無論如何，在延誤的原因出現後28天內以書面通知「工程師」延誤的原因及可能延誤的程度。

但是，當「承建商」能合理地預見「工程師」發出的任何命令或指令有可能引致「本工程」或其任何「工段」的進展受到延誤，「承建商」必須立即以書面通知「工程師」，及說明此項延誤可能造成的影響以及可能延誤的程度。在任何情況下，該通知書必須在不遲於「工程師」發出有關命令或指令後28天內發出。

(b) 如「工程師」認為此延誤的原因是

(i) 惡劣天氣及/或其後果對「本工程」的進展有負面影響；或

(ii) 懸掛八號或更高的熱帶風暴警告；或

(iia) 黑色暴雨警告；或

(iii) 「工程師」根據第5條發出的指令；或

(iv) 根據第60條發出的變更命令；或

(v) 在「合同」內的任何工程項目的數量有頗大幅度增加而不是根據第60條發出的變更命令所引致的；或

(vi) 「僱主」沒有根據「合同」給予「承建商」的「工地」或其任何分區或部分的管用權，或之後將「承建商」此等管用權剝奪；或

(vii) 「本工程」的進展受到干擾而責任在於「僱主」或「工程師」或「專門承建商」，包括但不限於第63條所提及的任何事宜；或

(viii)「工程師」根據第54條擱置「本工程」，但擱置並非由於第54(2)(a)至(d)條所敘述的情況所引致；或

(ix) 任何公用事業機構或其他依法組成的當局未能依時開始或施行任何直接影響「本工程」實施的工程，而「承建商」須已採取一切實際措施以令該公用事業機構或依法組成的當局開始或施行該等工程；或

(x) 任何「指定次承判商」因為本條第(b)(i)至(ix)款所述的任何原因引致延誤，而「承建商」已採取一切合理措施避免或減少延誤；或

(xi) 無論屬何種類的特殊情況，

則「工程師」必須在合理時間內考慮「承建商」是否公道地享有「本工程」或其任何「工段」竣工限期的延長。

Commentary

For most construction contracts, the time or times for completion of the Works counting from the date for commencement are specified. If there is any delay to the completion of the Works or any Section of it for which the Contractor is not responsible, this Clause allows extension of time to be granted to the Contractor by the Engineer. The list of events entitling extension of time covers: (1) delay which is the responsibility of the Employer or the Engineer as his agent such as clarification of Contract documents by the Engineer, variations, late information from the Engineer, increase in quantity of work, delay in possession of the Site, Engineer's instruction causing delay and suspension as given in (iii) to (viii) and (x) of the list, and, (2) delay which is caused by natural events such as inclement weather, utility undertakings and special circumstances given in the rest of the list. At first sight, this Clause appears to benefit the Contractor only but in fact it is also a very important provision for the Employer to prevent time being rendered at large due to Employer's act of prevention,[1] such as failure to give possession of the Site or late instruction. This Clause also describes the notice requirement which the Contractor needs to comply with. After giving notice, the Contractor needs to submit full particulars to the Engineer to justify his claim for extension of time. The Engineer is then required to consider all circumstances known to him at that time when assessing extension of time, including the effect of omission of work or substantial decrease in quantities. After having considered the Contractor's application, the Engineer has to notify the Contractor any extension of time granted, or if he is not entitled to any, within a reasonable time so that the Contractor can plan his work properly. The last important point in this Clause is that there is no compensation of associated cost in connection with the extension of time granted, unless there are other provisions elsewhere in the Contract which state otherwise.

Analysis and applications

Clause 50(1)(a) "As soon as practicable but in any event within 28 days after the cause of any delay to the progress of the Works or any Section thereof has arisen, the Contractor shall give notice …"

The Clause essentially says that (1) notice should be given as soon as practicable, e.g., any delay by utilities undertaking's failure to carry out diversion should be notified to the Employer preferably on the day when the delay occurs or the next day or so, (2) the maximum notice period is 28 days. Early notice is important because the Engineer or the Employer may implement mitigation measures or may oblige/remind the Contractor to fulfil his general duty to mitigate the delay. Depending on the circumstance of delay, the Contractor's failure to give notice "as soon as practicable" well before the 28 days may be taken into account when he grants the extension of time (see 1st proviso to sub-clause (2), which requires the Engineer to take into account all the circumstances and the Contractor's delay to give timely notice may be regarded as a circumstance). Another point to consider is the effect of the Contractor's failure to give notice within 28 days. Does it bar the Contractor's claim for extension of time? Decided cases say that notice within the 28 days shall not be a condition precedent.[2] The practical effect is that failure to give notice within the 28 days stipulated will at most prejudice the claim for extension of time, e.g., when the relevant records are lost or that the facts cannot be recalled.

Clause 50(1)(a) "… Provided that as soon as the Contractor can reasonably foresee that any order or instruction issued by the Engineer is likely to cause a delay to the progress of Works … shall forthwith give notice … Such notice shall not in any event be given later than 28 days …"

Examples of such order or instruction are variation orders, suspension orders, or orders to open up for inspection. The most common ones are variation orders. If an order is given, the Contractor must foresee whether there is any delay as far as reasonable. He shall "forthwith" give notice to the Engineer of any possible delay in order that the Engineer may implement mitigation measures or even withdraw the instruction if the anticipated delay by the Contractor is excessive. Again, failure of the Contractor to give notice "forthwith" or in any case within 28 days may prejudice the Contractor's claim for extension of time.

Clause 50(1)(b) "If in the opinion of the Engineer the cause of delay is: (ix) any utility undertaking or other duly constituted authority failing to commence or to carry out in due time any work directly affecting the execution of the Works, provided that the Contractor has taken all practical steps to cause the utility undertaking … to commence or proceed …"

This Clause operates only when there is failure on the part of utility undertaking or other duly constituted authority, e.g., the MTRC's diversion of water-cooling mains, to commence or to carry out the diversion works. It does not operate when, say, there is an uncharted watermain. The sub-clause also requires the Contractor to take all practical steps to cause the utility undertaking, etc., to commence or to proceed with their works. To satisfy such requirements, the Contractor needs to arrange meetings to resolve conflicts among the utility undertakings, arrange working schedule, confirm the time when the Site is available and offer reasonable facilities.

Clause 50(1)(b)(xi) "any special circumstance of any kind whatsoever."

This is a very wide sub-clause which generally covers events which are exceptional and out of the ordinary. There have been judicial cases, not in civil engineering context, which considered the words "special circumstances" and all have concluded that the wording refers to something out of the ordinary.[3] A decided case suggests that a special circumstance is a circumstance outside the control or contemplation of either party.[4] Also, Professor Uff's commentary on the ICE Conditions in *Keating on Building Contracts*[5] stated that "matters reasonably within the contemplation of the parties, such as delay by the sub-contractor (whether nominated or not) will not rank as special." It is submitted that while delay by sub-contractors or suppliers (not Nominated Sub-contractor which is allowed for in items (x) of the list) generally does not entitle the Contractor to extension of time since the sub-contractors or suppliers are normally (but not always, depending on the facts of the case[6]) within the control of the Contractor, not everything within the contemplation of the parties should be excluded from being considered as a special circumstance. Situations like worldwide shortage of a particular type of construction material may be considered as a special circumstance. Even uncharted utilities or unforeseen adverse ground conditions are sometimes considered by Engineer as special circumstances.

Clause 50 "… then the Engineer shall within a reasonable time consider whether the Contractor is fairly entitled to an extension of time …"

Many engineers adopt a "wait and see" attitude when considering extension of time in order to avoid granting extension of time which cannot be reduced subsequently. In many cases, the Engineer may blame the Contractor for not submitting full particulars and the process can drag on for a long time. It is submitted that a proper way to consider extension of time is to grant the Contractor extension of time in stages based on information available in accordance with sub-clause (3). It is then up to the Contractor to provide more information and request for a review under the 2nd proviso to sub-clause (2). The wording "fairly entitled" may suggest the following considerations. First of all, the delay event must affect critical activities on the critical path, which has an effect on the dates of completion. Secondly, if there are two delay events (concurrent delay) and one of which is due to the Contractor's own fault, i.e., not a relevant event, and the other which the Contractor may have entitlement, i.e., Employer's delay or neutral delay, should extension of time be granted? The current thinking as expressed in SCL Protocol is positive.[7] The principle is that the Employer should not take advantage of the Contractor's delay to deprive the Contractor of any extension of time the Contractor may be entitled so as to impose liquidated damages on the Contractor. Thirdly, if there is some float, whom does the float belong to? The SCL Protocol suggests the float belongs to the project, not to either Party. Though it appears whoever uses the float first should have it, the float should be used in a reasonable manner. Fourthly, consideration has to be given to the Contractor's general duty to mitigate the delay using at least reasonable endeavour[8] which means rearranging resources and working arrangement, etc., but not requiring significant additional resources and costs.

 The Engineer has approved the location of this gantry for lifting precast viaduct segments. However, the Marine Department now restricts the loading activities to non-peak hours because of heavy marine traffic. What are the Contractor's remedies?

Clause 50 (continued)

Extension of Time for Completion

(1) (c) Notwithstanding the powers of the Engineer under the provisions of this Clause to decide whether the Contractor is fairly entitled to an extension of time the Contractor shall not be entitled to an extension of time for the completion of the Works or any Section thereof if the cause of the delay is:

    (i) a suspension occasioned by the circumstances described in Clause 54(2)(a) to (d), or

    (ii) a shortage of Constructional Plant or labour.

(2) If in accordance with sub-clause (1) of this Clause the Engineer considers that the Contractor is fairly entitled to an extension of time for the completion of the Works or any Section thereof, the Engineer shall within a reasonable time determine, grant and notify in writing to the Contractor such extension. If the Engineer decides that the Contractor is not entitled to an extension, the Engineer shall notify the Contractor in writing accordingly.

    Provided that the Engineer in determining any such extension shall take into account all the circumstances known to him at that time, including the effect of any omission of work or substantial decrease in the quantity of any item of work.

    Provided further that the Engineer shall, if the Contractor shall so request in writing, make a subsequent review of the circumstances causing delay and determine whether any further extension of time for completion should be granted.

(2A) For the avoidance of doubt if the Engineer grants an extension of time in respect of a cause of delay occurring after the Employer is entitled to recover liquidated damages in respect of the Works or any Section, the period of extension of time granted shall be added to the prescribed time or previously extended time for the completion of the Works or, as the case may be, the relevant Section.

(3) For the purposes of determining whether or to what extent the Contractor may be entitled to an extension of time under sub-clause (1)(b) of this Clause the Engineer may require the Contractor to submit full and detailed particulars of the cause and extent of the delay to the progress of the Works. Where such full and detailed particulars are required by the Engineer, they shall be submitted in writing by the Contractor to the Engineer as soon as practicable in order that the Contractor's claim may be investigated at that time by the Engineer. If the Contractor fails to comply with the provisions of this sub-clause, the Engineer shall consider such extension only to the extent that the Engineer is able on the information available.

(4) Whenever the Engineer grants an extension of time for completion in accordance with this Clause, the Contractor shall revise the programme referred to in Clause 16 accordingly.

(5) Except as provided elsewhere in the Contract, any extension of time granted by the Engineer to the Contractor shall be deemed to be in full compensation and satisfaction for any loss or injury sustained or sustainable by the Contractor in respect of any matter or thing in connection with which such extension shall have been granted and every extension shall exonerate the Contractor from any claim or demand on the part of the Employer for the delay during the period of such extension but not for any delay continued beyond such period.

(6) For the purpose of this Clause, "Black Rainstorm Warning" means a warning issued by the Director of the Hong Kong Observatory of a heavy rainstorm in, or in the vicinity of, Hong Kong by the use of the heavy rainstorm signal commonly referred to as Black.

Corresponding clauses in the ICE Conditions:

Clause 44 (*continued*)

(3) Should the Engineer consider that the delay suffered fairly entitles the Contractor to an extension of the time for the substantial completion of the Works or any Section thereof such interim extension shall be granted forthwith and be notified to the Contractor in writing with a copy to the Employer. In the event that the Contractor has made a claim for an extension of time but the Engineer does not consider the Contractor entitled to an extension of time he shall so inform the Contractor without delay.

(4) The Engineer shall not later than 14 days after the due date or extended date for completion of the Works or any Section thereof (and whether or not the Contractor shall have made any claim for an extension of time) consider all the circumstances known to him at that time and take action similar to that provided for in sub-clause (3) of this Clause. Should the Engineer consider that the Contractor is not entitled to an extension of time he shall so notify the Employer and the Contractor.

Corresponding clauses in the FIDIC Conditions:

Clause 8.4 (*continued*)

If the Contractor considers himself to be entitled to an extension of the Time for Completion, the Contractor shall give notice to the Engineer in accordance with Sub-Clause 20.1 [*Contractor's Claims*]. When determining each extension of time under Sub-Clause 20.1, the Engineer shall review previous determinations and may increase, but shall not decrease, the total extension of time.

Clause 8.5

If the following conditions apply, namely:

(a) the Contractor has diligently followed the procedures laid down by the relevant legally constituted public authorities in the Country,

(b) these authorities delay or disrupt the Contractor's work, and

(c) the delay or disruption was unforeseeable,

then this delay or disruption will be considered as a cause of delay under sub-paragraph (b) of Sub-Clause 8.4

條款 50

香港特別行政區土木工程合同一般條件：

延長竣工限期

（續）

(1) (c) 雖然「工程師」根據本條文有權決定「承建商」是否可公道地享有竣工限期的延長，但如延誤的原因是：

    (i) 第54(2)(a)至(d)條所敘述的情況所引致工程的擱置；或

    (ii) 「施工設備」或勞工短缺，

    則「承建商」不可享有「本工程」或其任何「工段」竣工限期的延長。

(2) 如果「工程師」根據本條第(1)款認為「承建商」可公道地享有「本工程」或其任何「工段」的竣工限期的延長，「工程師」必須在合理時間內決定、給予並以書面通知「承建商」該等完工限期的延長。如「工程師」決定「承建商」不能享有竣工限期的延長，他必須就此以書面通知「承建商」。

    但「工程師」為那些竣工限期的延長作決定時，必須考慮當時他所知道的一切情況，包括任何被刪除的工程或任何工程項目數量有頗大幅度減少所引致的影響。

    再者，如「承建商」以書面提出要求，「工程師」必須覆核延誤的情況，並且決定應否給予多一些竣工限期的延長。

(2A) 為避免產生疑問，如「工程師」因為任何延誤的原因給予竣工限期的延長，而該原因是在「僱主」有權追討「本工程」或其任何「工段」的經算定賠償之後才出現的，則所給予的該段竣工限期的延長必須加於「本工程」或有關「工段」（視情況而定）所訂定竣工限期或先前已延長的竣工限期之上。

(3) 為決定「承建商」按照本條第(1)(b)款可否享有竣工限期的延長及其程度，「工程師」可要求「承建商」提交「本工程」進展延誤的原因及程度的全部詳情。如「工程師」有此要求，「承建商」必須在切實可行的情況下盡快以書面提交全部詳情予「工程師」，以便「工程師」可在當時調查「承建商」的索償。如「承建商」未能遵從本款的條文，「工程師」只須根據他所得到的資料在所能範圍內考慮該等竣工限期的延長。

(4) 每當「工程師」根據本條給予「承建商」竣工限期的延長,「承建商」必須就此修改第16條所提及的施工計劃。

(5) 除本「合同」另有規定外,「工程師」所給予「承建商」的任何竣工限期的延長必須視為「承建商」因任何事情或事宜而遭受或可能遭受的任何損失或人身損傷的全部補償及已使「承建商」滿意;而且,上述每次的延長皆免除「承建商」受到來自「僱主」因在此期間延誤的任何索償及要求,但超過此期間的持續延誤則除外。

(6) 為本條的作用而言,「黑色暴雨警告」是指由香港天文台台長使用通常被稱為「黑色」的暴雨警告訊號而發出的有關在香港或香港附近出現暴雨的警告。

Analysis and application

Clause 50(1)(c) "… the contractor shall not be entitled to an extension of time … if the cause of delay is: (i) a suspension occasioned by the circumstances described in Clause 54(2)(a) to (d), or (ii) a shortage of Construction Plant or labour."

The sub-clause is quite self-explanatory but it should be noted that shortage of materials is not included.

Clause 50(2) "… within a reasonable time determine, grant and notify in writing to the Contractor such extension …"

As discussed in the commentary, the Engineer should not deliberately postpone the granting of extension of time because in doing so, there is the danger that the Contractor may succeed in claiming for acceleration costs. A recent English case decided that if the Contractor is entitled to extension of time arising from an Employer's delay event but is denied by the Engineer and the Contractor spends extra resources to accelerate the works in order to avoid the imposition of liquidated damages, then the Contractor is entitled to the acceleration costs.[9]

Clause 50(2) "… Provided that the Engineer in determining any such extension [of time] shall take into account all the circumstances known to him at the time, including the effect of any omission of work or substantial decrease in the quantity of any item of work …"

The wording "all the circumstances known to him at the time" is worth discussing in some detail. The Engineer shall consider whether the Contractor is "fairly entitled" to any extension of time by referring to these circumstances. Apart from considering critical path, concurrent delay, floats and duty to mitigate as discussed above in the commentary, it is submitted that the manner of the Contractor in carrying out the work, such as whether he carries out the work with due diligence, his efficiency, and whether he follows his original plan, etc., should be taken into account. For example, if the Contractor postpones carrying out the work until a time that the work becomes critical, i.e., using up the float in an unreasonable manner, it can be taken into account even if the subsequent delay event causes actual delay to the Works. Another example is that if the Engineer orders variations to avoid delay in diversion of utilities for which the Contractor is responsible, the Engineer should take into account the original timing necessary to complete the Works had the variations not been ordered.

Clause 50(2A) "For the avoidance of doubt if the Engineer grants an extension of time in respect of a cause of delay occurring after the Employer is entitled to recover liquidated damages … the period of extension of time granted shall be added to the prescribed time or previously extended time for the completion of the Works …"

This sub-clause means that if, for example, the Contractor has not yet completed the Works on 28 December 2005, which is beyond the due date for completion, i.e., 1 December 2005, and he is therefore liable to liquidated damages. However, during that period, if there is any delay event to which he is entitled to an extension of time of, say 5 days, his revised due date or extended date of completion will then be 6 December 2005.

Clause 50(3) "... submit full and detailed particulars of the cause and extent of the delay to the progress of the Works ... If the Contractor fails to comply with the provisions of this sub-clause, the Engineer shall consider such extension only to the extent that the Engineer is able on the information available."

It is important to remind the Contractor in writing to submit full and detailed particulars after acknowledging the Contractor's notice of claim, quoting this sub-clause. The wording "shall consider such extension only to the extent that the Engineer is able on the information available" puts an obligation on the Engineer to grant extension of time even full particulars have not yet been submitted. As stated before, if not all information are available, or that the delay event is continuing, the Engineer should grant extension of time in stages as far as possible to avoid facing acceleration claims. Apart from this, proactive assessment of extension of time as the delay event arises may obviate the need to employ complicated delay analysis techniques to assess the extension of time after the construction is completed. However, it is desirable for the Engineer to familiarize himself with these techniques such as "what if (as planned impact)" analysis, "but for (as-built collapse)" analysis and time slice (Window) analysis. Of these, the time slice analysis is favoured by most practitioners involved since it takes into account the relevant programmes and the actual progress at the relevant times. These methods are also recommended in the SCL Protocol.

Clause 50(4) "Whenever the Engineer grants an extension of time ..., the Contractor shall revise the programme referred to in Clause 16 accordingly."

It should be noted that the revision of programme, especially for major projects consisting of thousands of activities, is not a simple task. The Engineer must refrain from exercising his power to require the Contractor to revise his programme too frequently, especially when extension of time for inclement weather is granted.

Clause 50(5) "Except as provided elsewhere in the Contract, any extension of time granted by the Engineer to the Contractor shall be deemed to be in full compensation and satisfaction for any loss or injury sustained or sustainable by the Contractor ..."

This sub-clause simply means that in granting extension of time under this Clause 50, no compensation for any loss or injury caused by the delay event shall be made by the Employer, unless another Clause in the HKSAR Conditions states otherwise, i.e., the Employer's delay event from items (iii) to (viii) of the list.

Comparison with the ICE Conditions

This ICE Clause 44 requires the Contractor to deliver to the Engineer full and detailed particulars of any extension time claimed within 28 days after the cause of any delay has arisen or as soon thereafter as is reasonable in order that the claim may be investigated at the time. There is not any notice requirement under this Clause. However, if the claim for extension of time also involves prolongation cost and other costs, the Contractor must give notice pursuant to Clause 53. Also, unlike the HKSAR Clause, there is no apparent requirement for the Contractor to reasonably foresee any order or instruction by the Engineer which may cause delay. One apparent advantage of the HKSAR Clause is that the Engineer may consider withdrawing or reconsidering a variation order if the Contractor claims a substantial extension of time, as explained earlier. One other difference is that one item of the ICE Clause refers to "any cause of delay referred to in these Conditions" but does not list them out. The relevant Clauses are Clause 12 on adverse physical conditions and artificial obstruction, Clause 13 on instructions, Clause 40 on suspension and Clause 42 on possession of the Site. The ICE Clause allows only for "exceptional" adverse weather conditions and the Contractor has to find out the hydrological conditions himself pursuant to Clause 11(2)(a) when pricing his bid. Sub-clause (1)(c) allows the Engineer to grant extension of time for any delay impediment prevention or default by the Employer to prevent time being set "at large." Sub-clause 2(b) enables the Engineer to grant extension of time even not claimed by the Contractor also to prevent time at large. The Government of the HKSAR has a similar sub-clause in the Special Conditions for Mega Project Contracts. Sub-clause (3) expressly empowers the Engineer to grant interim extension of time where sub-clause (4) obligates the Engineer to review the overall extension of time within 14 days after the due date or extended date for completion of the Works. Sub-clause (5) (not shown here) obligates the Engineer to make a review within

28 days of the issue of Certificate of Substantial Completion and make a final determination of extension of time. This ICE Clause imposes a duty on the Engineer to grant extension of time in a number of stages and this is good because it enables the Contractor to know his position well and plan his work properly.

Comparison with the FIDIC Conditions

In this FIDIC Clause, the wording "to the extent that completion for the purpose of Sub-Clause 10.1 [*Taking over of the Works and Sections*] is or will be delayed by any of the following cause" suggests that the delay must be affecting critical activities leading to the taking over of the Works and Sections before giving rise to an entitlement to any extension of time. The items in the list are similar to those of the ICE Conditions except that the FIDIC Conditions cover "unforeseeable shortages in the availability of personnel or Goods caused by epidemic or governmental actions." It should be noted that shortage of personnel due to widespread disease or change in policy on labour importation, etc., may entitle the Contractor to an extension of time. In the HKSAR Conditions and the ICE Conditions, these may be covered by special circumstances. On the other hand, the FIDIC Clause does not have an item for extension of time arising from special circumstances though Clause 19 on Force Majeure allows an event beyond the Party's control, not foreseeable and not surmountable to excuse the Contractor from any future performance. Like the HKSAR Clause, the Contractor may have an entitlement to extension of time if the delay is caused by the duly constituted authority on the condition that the Contractor has diligently followed the laid-down procedures. Lastly, the FIDIC Clause allows the Engineer to review any previous determination when the Contractor gives notices to the Engineer in accordance with sub-clause 20.1, and the review may increase but not reduce the total extension of time.

Notes

1. *Holme v Guppy* (1838) 3M & W 387; *Wells v Army Navy Co-operative Society Ltd.* (1902) *Hudson Building Cases*, 4th Edition, Vol. 2, p. 346.

2. *Bremer Handelsgesellschaft v Vanden Avenue-Izegem* (1978) 2LLR109

3. *Cortex Investments v Olphert v Collins* [1984] 2NZLR; *Lyon v Eilcox* [1994] 3NZLR422; *The Expile Property Ltd. v Jabb's Excavations Property Ltd.* (2002) 194ALR.

4. *Wells v Army Navy Co-operative Society Ltd.* (1902) *Hudson Building Cases*, 4th Edition, Vol. 2, p. 346.

5. Vivian Ramsey and Stephen Furst, *Keating on Building Contracts*, 6th Edition.

6. *Scott Lithgow v Secretary of State for Defence* (1989) 45BLR6 held that failure of suppliers or sub-contractors in breach of their contractual obligation were not matters which, according to the ordinary use of language, could be regarded as being within the Contractor's control.

7. See also Roger Knowles, *150 Contractual Problems and Their Solutions*, Section 3.5.

8. For a comparison between the best endeavour and reasonable endeavour, see Roger Knowles, *150 Contractual Problems and Their Solutions*, Section 3.4.

9. *Motherwell Bridge Construction v Micafil Vakuumterhnik and Anothers* (2002) TCC81CONLR44.

Clause 51

Rate of Progress

(1) If the rate of progress of the Works or any Section thereof is at any time in the opinion of the Engineer too slow to ensure completion by the prescribed time or extended time for completion, the Engineer may so inform the Contractor in writing and the Contractor shall immediately take such steps as are necessary to expedite the completion of the Works or any Section thereof. The Contractor shall inform the Engineer of such proposed steps and revise the programme referred to in Clause 16 accordingly.

(2) Notwithstanding the provisions of sub-clause (1) of this Clause and subject to compliance with any enactment, regulation or bye-law, the Engineer shall be empowered to instruct the Contractor in writing to carry out the Works or any part thereof during any hours of the day where the Engineer considers it necessary owing to the default, negligence, omission or slow progress of the Contractor.

(3) The Contractor shall not be entitled to any additional payment for complying with any instruction given in accordance with this Clause.

Corresponding clauses in the ICE Conditions:

Clause 46

(1) If for any reason which does not entitle the Contractor to an extension of time the rate of progress of the Works or any Section is at any time in the opinion of the Engineer too slow to ensure substantial completion by the time or extended time for completion prescribed by Clauses 43 and 44 as appropriate or the revised time for completion agreed under sub-clause (3) of this Clause the Engineer shall notify the Contractor in writing and the Contractor shall thereupon take such steps as are necessary and to which the Engineer may consent to expedite the progress so as substantially to complete the Works or such Section by that prescribed time or extended time. The Contractor shall not be entitled to any additional payment for taking such steps.

(2) If as a result of any notice given by the Engineer under sub-clause (1) of this Clause the Contractor seeks the Engineer's permission to do any work on Site at night or on Sundays such permission shall not be unreasonably refused.

(3) If the Contractor is requested by the Employer or the Engineer to complete the Works or any Section within a revised time being less than the time or extended time for completion prescribed by Clauses 43 and 44 as appropriate and the Contractor agrees so to do then any special terms and conditions of payment shall be agreed between the Contractor and the Employer before any such action is taken.

Corresponding clauses in the FIDIC Conditions:

Clause 8.6

If, at any time:

(a) actual progress is too slow to complete within the Time for Completion, and/or

(b) progress has fallen (or will fall) behind the current programme under Sub-Clause 8.3 [*Programme*],

other than as a result of a cause listed in Sub-Clause 8.4 [*Extension of Time for Completion*], then the Engineer may instruct the Contractor to submit, under Sub-Clause 8.3 [*Programme*], a revised programme and supporting report describing the revised methods which the Contractor proposes to adopt in order to expedite progress and complete within the Time for Completion.

Unless the Engineer notified otherwise, the Contractor shall adopt these revised methods, which may require increases in the working hours and/or in the numbers of Contractor's Personnel and/or Goods, at the risk and Cost of the Contractor. If these revised methods cause the Employer to incur additional costs, the Contractor shall subject to Sub-Clause 2.5 [*Employer's Claims*] pay these costs to the Employer, in addition to delay damages (if any) under Sub-Clause 8.7 below.

條款 51 進展

(1) 在任何時候，如「工程師」認為「本工程」或其任何「工段」的進展過慢，使到不能保証「本工程」或其任何「工段」在訂定竣工限期或已延長的竣工限期內完成，「工程師」可以就此以書面通知「承建商」，而「承建商」必須立刻採取所需措施加快完成「本工程」或其任何「工段」。「承建商」必須將所建議的措施通知「工程師」，並就此相應地修改第16條所提及的施工計劃。

(2) 雖然本條第(1)款的條文有所規定，但如「工程師」認為由於「承建商」的失責、疏忽、刪減工作或進展過慢而有需要，在符合任何法例、規例或附例的規定下，「工程師」有權以書面指令「承建商」在每天的任何時間進行「本工程」或其任何部分的工作。

(3) 「承建商」不會因依從根據本條發出的任何指令而獲得額外付款。

Commentary

This Clause is useful in the sense that if the Contractor falls behind the programme such that he may not be able to complete on time, the Engineer may invoke this Clause to require the Contractor to take immediate steps to expedite the completion of the Works. The use of this preventive power is at the discretion of the Engineer and would be welcomed by the Employer because the latter usually wants the Works to be completed on time and not the liquidated damages. Before the use of this power, the Engineer must consider whether there are grounds for further extension of time. The Clause also empowers the Engineer to instruct the Contractor to carry out evening or night work if the Engineer considers it necessary owing to default, negligence or slow progress of the Contractor. For example, if the falsework of a critical bridge span is incorrectly erected and the following concreting operation involves pre-arranged traffic diversion, the Engineer may instruct night work on rectifying the falsework in order that concreting can be carried out in the pre-arranged time subject to complying with the law such as obtaining night works permit. Such exercise of the above powers by the Engineer shall not entitle the Contractor to claim any extra costs.

Analysis and application

Clause 51(1) "… the Engineer may so inform the Contractor in writing and the Contractor shall immediately take such steps as are necessary to expedite the completion of the Works …"

It should be noted that there are very limited sanctions, which would be available if the Contractor is not taking steps to expedite the completion after being informed by the Engineer of the slow progress. Forfeiture under Clause 81 and carrying out of the Works by persons other than the Contractor under Clause 82 may both not be practical and even if it is unavoidable, the Engineer must build up sufficient evidence of the Contractor's failure to comply with his orders before taking such drastic actions. Very often, the only sanction is continuous warnings and giving bad performance reports.

Clause 51(2) "… subject to compliance with any enactment, regulation or bye-law, the Engineer shall be empowered to instruct the Contractor in writing to carry out the Works or any part thereof during any hours of the day where the Engineer considers it necessary owing to the default, negligence, omission or slow progress of the Contractor."

If the Engineer wishes to instruct night works owing to the slow progress of the Contractor, he has to make sure that he has given full consideration to any extension of time which the Contractor may be entitled to and grant such extension of time, otherwise his instruction may be exposed to claim by the Contractor. It appears, however, that if the instruction to carry out night work is "owing to the default, negligence, omission or slow progress" of the Contractor, say to complete any remedial works for the timely commencement of another major critical activity, the instruction would not attract any claim even though there is no evidence to support that the Contractor cannot complete on time.

Comparison with the ICE Conditions

This ICE Clause 46 is preceded by Clause 45, which prohibits any works during night-time and Sunday and unlike the HKSAR Clause, the ICE Clause is only applicable if in the opinion of the Engineer, the progress of the Works is too slow to ensure timely completion. Sub-clause (2) almost confers a right to the Contractor to carry out night work because the Engineer shall not unreasonably refuse to consent to the Contractor's proposal to carry out night work in order to ensure completion by the time or extended time for completion. Sub-clause (3) is effectively an agreement between the Contractor and the Employer on any special terms for acceleration of the Works.

Comparison with the FIDIC Conditions

The FIDIC Clause 8.6 differs from the HKSAR Clause and the ICE Clause in the sense that its applicability extends to the situation when progress has fallen behind the current programme, and not necessarily limited to the situation when the progress is too slow to complete within the time for completion. Upon instruction of the Engineer, the Contractor has to submit a revised programme and the revised methods to ensure timely completion. Such revised methods may require increase in working hours, the Contractor's personnel and goods. A major feature of this FIDIC Clause is that if the Contractor has to increase working hours or revise methods, the Employer can claim the Contractor the cost associated with the extra supervisory staff hours or the Engineer's time in approving the revised methods, etc.

 23 The progress of the flyover construction is seriously lagging behind programme. The Engineer instructs night works. What should the Engineer consider before giving such an instruction? Can the Contractor claim additional payment?

This is a specially designed moving platform for the construction of a prestressed in situ variable cross-sectional bridge. The progress is lagging behind the programme. Can the Engineer instruct the Contractor to procure one more such platform to expedite the progress? What is the Employer's liability if the extra platform collapses?

Clause 52

Liquidated Damages for Delay

(1) If the Contractor fails to complete the Works or where the Works are divided into Sections any Section within the time for completion prescribed by Clause 49 or such extended time as may be granted in accordance with Clause 50, then the Employer shall be entitled to recover from the Contractor liquidated damages, and may but shall not be bound to deduct such damages either in whole or in part, in accordance with the provisions of Clause 83. The payment of such damages shall not relieve the Contractor from his obligations to complete the Works or from any other of his obligations under the Contract.

(2) The liquidated damages shall be calculated using the rate per day prescribed in the Appendix to the Form of Tender, either for the Works or for the relevant Section, whichever is applicable.

Provided that, if the Engineer certifies completion under Clause 53 of any part of the Works before completion of the Works or any part of any Section before the completion of the whole thereof, then the rate per day of liquidated damages for the Works or the relevant Section shall from the date of such certification be reduced in the proportion which the value of the part so certified bears to the value of the Works or the relevant Section, as applicable, both values as of the date of such certification shall be determined by the Engineer.

Corresponding clauses in the ICE Conditions:

Clause 47

(1) (a) Where the whole of the Works is not divided into Sections the Appendix to the Form of Tender shall include a sum which represents the Employer's genuine pre-estimate (expressed per week or per day as the case may be) of the damages likely to be suffered by him if the whole of the Works is not substantially completed within the time prescribed by Clause 43 or by an extension thereof granted under Clause 44 or by any revision thereof agreed under Clause 46(3) as the case may be.

(b) If the Contractor fails to achieve substantial completion of the whole of the Works within the time so prescribed he shall pay to the Employer the said sum for every week or day (as the case may be) which shall elapse between the date on which the prescribed time expired and the date the whole of the Works is substantially completed.

Provided that if any part of the Works is certified as substantially complete pursuant to Clause 48 before the completion of the whole of the Works the said sum shall be reduced by the proportion which the value of the part so completed bears to the value of the whole of the Works.

(2) (a) Where the Works is divided into Sections (together comprising the whole of the Works) which are required to be completed within particular times as stated in the Appendix to the Form of Tender sub-clause (1) of this Clause shall not apply and the said Appendix shall include a sum in respect of each Section which represents the Employer's genuine pre-estimate (expressed per week or per day as the case may be) of the damages likely to be suffered by him if that Section is not substantially completed within the time prescribed by Clause 43 or by any extension thereof granted under Clause 44 or by any revision thereof agreed under Clause 46(3) as the case may be.

(b) If the Contractor fails to achieve substantial completion of any Section within the time so prescribed he shall pay to the Employer the appropriate stated sum for every week or day (as the case may be) which shall elapse between the date on which the prescribed time expired and the date of substantial completion of that Section.

Provided that if any part of that Section is certified as substantially complete pursuant to Clause 48 before the completion of the whole thereof the appropriate stated sum shall be reduced by the proportion which the value of the part so completed bears to the value of the whole of that Section.

Corresponding clauses in the FIDIC Conditions:

Clause 8.7

If the Contractor fails to comply with Sub-Clause 8.2 [*Time for Completion*], the Contractor shall subject to Sub-Clause 2.5 [*Employer's Claims*] pay delay damages to the Employer for this default. These delay damages shall be the sum stated in the Appendix to Tender, which shall be paid for every day which shall elapse between the relevant Time for Completion and the date stated in the Taking-Over Certificate. However, the total amount due under this Sub-Clause shall not exceed the maximum amount of delay damages (if any) stated in the Appendix to Tender.

These delay damages shall be the only damages due from the Contractor for such default, other than in the event of termination under Sub-Clause 15.2 [*Termination by Employer*] prior to completion of the Works. These damages shall not relieve the Contractor from his obligation to complete the Works, or from any other duties, obligations or responsibilities which he may have under the Contract.

條款 52

香港特別行政區土木工程合同一般條件：
對「本工程」延遲的算定賠償

(1) 如「承建商」未能按照第49條所訂定的竣工限期或根據第50條給予的延長限期內完成「本工程」或在「本工程」分為多個「工段」的情況下完成任何「工段」，則「僱主」可向「承建商」追討經算定的賠償。「僱主」可以但並非必須根據第83條的條文扣除整項或部分賠償。縱使「承建商」已支付該等賠償，亦不能免除他完成「本工程」的責任或他在「合同」下的任何其他責任。

(2) 計算經算定的賠償，必須採用在「投標表格」「附件」內為「本工程」或有關「工段」而訂定的每天價率（視乎那個項目適用）。

但是，如「工程師」根據第53條在「本工程」完成前核實「本工程」的任何部分為已完成，或在任何「工段」全部完成前核實該「工段」的任何部分為已完成，則「本工程」或有關「工段」經算定的賠償應由上述核實的日期翌日起計，而每天價率則必須按照經上述所核實的部分的價值佔「本工程」或有關「工段」（視乎那個項目適用）的價值的比例相應減少。此等價值必須由「工程師」釐定，及以上述核實日期那日計算的價值為準。

Commentary

This Clause is a common type of liquidated damages Clause entitling the Employer to damages for delay in completion. The Court recognizes that the parties to a contract may like to agree in advance the amount of compensation to be paid in case of any breach, rather than leaving it to be assessed by the Court because in some complex contracts, the damages to a party are extremely difficult to prove. With liquidated damages agreed beforehand when entering into the Contract, the amount of compensation is automatically fixed once the length of delay is determined and the actual amount suffered by a Party is irrelevant. However, many disputes arise because the Contractor liable to pay liquidated damages often challenges that the liquidated damages are actually penalty, and the latter is likely to be struck down by the Court. Recent judgments show that the Court tends to uphold the liquidated damages clause provided that the liquidated damages are genuine pre-estimate of the damages.

Analysis and application

Clause 52(1) "If the Contractor fails to complete the Works ... within the time for completion ... or such extended time ..., then the Employer shall be entitled to recover from the Contractor liquidated damages, ... but shall not be bound to deduct such damages ..."

Apparently, the application of liquidated damages is straightforward but actually there are a number of considerations. First of all, it is necessary to determine the contractual time for completion, which can only be finalized after all applications by the Contractor for extension of time are determined. Even these are determined by the Engineer, the total extension of time may be disputed and fall to be settled by mediation or arbitration. Secondly, it is necessary to determine the actual completion date, which depends on the substantial completion of the Works. What constitutes substantial completion is also not a straightforward determination as explained later in the commentary to Clause 53. Apart from the above, there is also a strategic consideration. The prime consideration of the Employer, and especially the Government, is to complete the Works within time, and not the deduction of liquidated damages. Usually, the issue of liquidated damages comes up at the critical times when the Contractor has to step up all his resources to complete the Works and it may not be wise for the Employer to hinder the cash flow and damage the relationship with the Contractor by deduction of liquidated damages. Moreover, many reputable contractors see the liquidated damages as a kind of penalty for failing to complete the Works and they will fight for every claim for extension of time, even though some of the claims are not supportable. For many Mega Project Contracts of contract sum in the order of billions of dollars, the amount of liquidated damages can be up to half a million Hong Kong dollars per day and it is not uncommon for the Contractor to challenge the basis of the pre-determined liquidated damages. As explained above in the commentary, the Court tends to uphold liquidated damages provided that it is a genuine pre-estimate. A Privy Council case[1] held that provided that the amount of liquidated damages is not extravagant or unconscionable, having regard to the range of losses that the Employer could reasonably anticipate that he has to suffer at the time of entering into the Contract, not at the time he suffers loss, then the liquidated damages will not be regarded as penalty. Even it is subsequently found that the liquidated damages considerably exceeds the foreseeable loss, say, due to errors in calculations, the Court still uphold the liquidated damages.[2] The Court also considers commercially justifiable reasons for upholding the validity of liquidated damages and not just the calculation.[3] Having said that, for a small contract which is on the critical path of a very large contract, it may not be appropriate simply to add on the anticipated damages arising from consequential delay of the large contract and any such calculations must be carefully considered. If liquidated damages are found to be invalidated, the Employer can claim general damages, which is subject to proof of damages. The wording "shall not be bound to deduct such damages" gives a free hand to the Employer to consider factors beyond financial considerations.

Clause 52(1) "... The payment of such damages shall not relieve the Contractor from his obligations to complete the Works or from any other of his obligations under the Contract."

In the HKSAR Government contracts, the amount of liquidated damages is calculated based on the loss of return on the capital, supervisory staff cost and fluctuations (all related to the estimated final Contract Cost) and any special damages like the actual cost of making alternative provision, interfacing contracts, etc. The obligation to complete the Works continues until completion which is the date when liquidated damages cease to be payable. The Contractor not only needs to complete the Works but also needs to use his best endeavour to complete on time despite his payment of liquidated damages or otherwise the Employer can invoke Clause 81 or 82, asking his own workers or other contractors to take over and complete the Works, or any part of the Works.[4]

Clause 52(2) "The liquidated damages shall be calculated using the rate per day prescribed in the Appendix to the Form of Tender ..."

For the current HKSAR Government contracts, the liquidated damages shown in the Appendix to the Form of Tender are in terms of a formula built up from the Cost of various components of the Works. The reason is that the Cost of the Contract is not known until the Tender is accepted. It is submitted that this is, strictly speaking, not quite necessary because some mere inaccuracy in calculations would not invalidate the liquidated damages.

Clause 52(2) "… Provided that, if the Engineer certifies completion under Clause 53 of any part of the Works before completion of the Works … the rate per day of liquidated damages … shall … be reduced in the proportion which the value of the part so certified bears to the value of the Works …"

This means when a part of the Works is substantially completed, the liquidated damages are only applicable to the remaining part of the Works. A proportion of the rate per day of liquidated damages reflecting that part of the Works is then deducted from the original rate. In the HKSAR contracts, a Special Condition provides that the reduction will not result in the rate per day of liquidated damages becoming smaller than the specified minimum, which usually reflects the minimum level of staff cost in supervising the remaining part of the Works, no matter how small the remaining part of the Works is.

For the HKSAR Mega Project Contracts or the past Airport Core Project Contracts, a certificate can be given for the achievement of a certain "Stage." After the completion of the columns for the bridge, can we allow the container activities passing between these columns?

Clause 52

Liquidated Damages for Delay

(continued)

(3) The period for which liquidated damages shall be calculated shall be the number of days from the prescribed date for completion or any extension thereof of the Works or the relevant Section until and including the certified date of completion.

Provided that, if the Engineer subsequently grants an extension of time which affects the period described above, then the Employer shall reimburse to the Contractor the liquidated damages for the number of days so affected at the rate described in sub-clause (2) of this Clause together with interest at the rate provided for in Clause 79(4) within 28 days of the granting of such extension of time.

(4) All monies payable by the Contractor to the Employer pursuant to this Clause shall be paid as liquidated damages for delay and not as a penalty.

Corresponding clauses in the ICE Conditions:

Clause 47 (*continued*)

(3) All sums payable by the Contractor to the Employer pursuant to this Clause shall be paid as liquidated damages for delay and not as a penalty.

(5) The Employer may

(a) deduct and retain the amount of any liquidated damages becoming due under the provision of this Clause from any sums due or which become due to the Contractor or

(b) require the Contractor to pay such amount to the Employer forthwith.

If upon a subsequent or final review of the circumstances causing delay the Engineer grants a relevant extension or further extension of time the Employer shall no longer be entitled to liquidated damages in respect of the period of such extension.

Any sum in respect of such period which may already have been recovered under this Clause shall be reimbursed forthwith to the Contractor together with interest compounded monthly at the rate provided for in Clause 60(7) from the date on which such sums were recovered from the Contractor.

Corresponding clauses in the FIDIC Conditions:

There is no corresponding clause.

香港特別行政區土木工程合同一般條件：

對「本工程」延遲的算定賠償

(3) 計算經算定的賠償的日數，是由「本工程」或有關「工段」的訂定竣工日期或任何已延長的竣工日期翌日起計，直至並包括經核實的竣工日期為止。

但是，如「工程師」日後給予的竣工限期的延長影響到上述的日數，「僱主」必須於作出給予該完工時限的延長後28天內，按本條第(2)款所述的價率為受影響的日數向「承建商」償還經算定的賠償及按第79(4)條規定的息率計算利息。

(4) 「承建商」按照本條須支付「僱主」的所有款項是對延誤的經算定的賠償而不是罰款。

Analysis and application

Clause 52(3)   The first paragraph of this sub-clause about calculation of liquidated damages is self-explanatory. The proviso to sub-clause (3) about over-deduction is also self-explanatory.

Clause 52(4)   "All monies payable by the Contractor to the Employer pursuant to this Clause shall be paid as liquidated damages for delay and not as a penalty."

The wording "not as a penalty" will not save a liquidated damages provision if the sum is found to be a penalty for being extravagant or unconscionable, as discussed before.

Comparison with tne ICE Conditions

There is no major difference between this ICE Clause 47 and the HKSAR Clause 52. The term "liquidated damages" is not used but instead it refers to as "a sum which represents the Employer's genuine pre-estimate … of the damages likely to be suffered by him."

Comparison with the FIDIC Conditions

The term "delay damages" is used. This FIDIC Clause 8.7 also specifies that delay damages are the Employer's Claims and subject to notice requirements. It also allows capping the delay damages to a maximum amount. The FIDIC Clause also explicitly says that delay damages does not apply to termination.

Notes

1. *Philips Hong Kong Ltd. v The Attorney General of Hong Kong* (1991) Privy Council Appeal No. 29.
2. *Alfred McAlpine Capital Projects v Tilebox Ltd.* [2005] EWHC 281.
3. *Murray v Leisureplay plc.* [2005] EWCA CIV963.
4. *Bath and North East Somerset District Council v Mowlem plc.* [2004] BLR153, CA. See also the commentaries to Clauses 81 and 82.

Clause 53

Completion of the Works

(1) When the Works have been substantially completed and have satisfactorily passed any final test that may be prescribed by the Contract, the Contractor may serve notice in writing to that effect to the Engineer, accompanied by an undertaking to carry out any outstanding work during the Maintenance Period, requesting the Engineer to issue a certificate of completion in respect of the Works. The Engineer shall, within 21 days of the date of receipt of such notice either:

(a) issue a certificate of completion stating the date on which, in the Engineer's opinion, the Works were substantially completed in accordance with the Contract and the Maintenance Period shall commence on the day following the date of completion stated in such certificate, or

(b) give instructions in writing to the Contractor specifying all the work which, in the Engineer's opinion, is required to be done by the Contractor before such certificate can be issued, in which case the Contractor shall not be permitted to make any further request for a certificate of completion and the provisions of sub-clause (2) of this Clause shall apply.

(2) Notwithstanding the provisions of sub-clause (1) of this Clause, as soon as in the opinion of the Engineer the Works have been substantially completed and satisfactorily passed any final test which may be prescribed by the Contract, the Engineer shall issue a certificate of completion in respect of the Works and the Maintenance Period shall commence on the day following the date of completion stated in such certificate.

Corresponding clauses in the ICE Conditions:

Clause 48

(1) When the Contractor considers that
   (a) the whole of the Works or
   (b) any Section in respect of which a separate time for completion is provided in the Appendix to the Form of Tender

has been substantially completed and has satisfactorily passed any final test that may be prescribed by the Contract he may give notice in writing to that effect to the Engineer or to the Engineer's Representative. Such notice shall be accompanied by an undertaking to finish any outstanding work in accordance with the provisions of Clause 49(1).

(2) The Engineer shall within 21 days of the date of delivery of such notice either
   (a) issue to the Contractor (with a copy to the Employer) a Certificate of Substantial Completion stating the date on which in his opinion the Works were or the Section was substantially completed in accordance with the Contract or

(b) give instructions in writing to the Contractor specifying all the work which in the Engineer's opinion requires to be done by the Contractor before the issue of such certificate.

If the Engineer gives such instructions the Contractor shall be entitled to receive a Certificate of Substantial Completion within 21 days of completion to the satisfaction of the Engineer of the work specified in the said instructions.

(3) (*Quoted on page 118*)

(4) If the Engineer considers that any part of the Works has been substantially completed and has passed any final test that may be prescribed by the Contract he may issue a Certificate of Substantial Completion in respect of that part of the Works before completion of the whole of the Works and upon the issue of such certificate the Contractor shall be deemed to have undertaken to complete any outstanding work in that part of the Works during the Defects Correction Period.

Corresponding clauses in the FIDIC Conditions:

Clause 10.1

Except as stated in Sub-Clause 9.4 [*Failure to Pass Tests on Completion*], the Works shall be taken over by the Employer when (i) the Works have been completed in accordance with the Contract, including the matters described in Sub-Clause 8.2 [*Time for Completion*] and except as allowed in sub-paragraph (a) below, and (ii) a Taking-Over Certificate for the Works has been issued, or is deemed to have been issued in accordance with this Sub-Clause.

The Contractor may apply notice to the Engineer for a Taking-Over Certificate not earlier than 14 days before the Works will, in the Contractor's opinion, be complete and ready for taking over. If the Works are divided into Sections, the Contractor may similarly apply for a Taking-Over Certificate for each Section.

The Engineer shall, within 28 days after receiving the Contractor's application:

(a) issue the Taking-Over Certificate to the Contractor, stating the date on which the Works or Section were completed in accordance with the Contract, except for any minor outstanding work and defects which will not substantially affect the use of the Works or Section for their intended purpose (either until or whilst this work is completed and these defects are remedied); or

(b) reject the application, giving reasons and specifying the work required to be done by the Contractor to enable the Taking-Over Certificate to be issued. The Contractor shall then complete this work before issuing a further notice under this Sub-Clause.

If the Engineer fails either to issue the Taking-Over Certificate or to reject the Contractor's application within the period of 28 days, and if the Works or Section (as the case may be) are substantially completed in accordance with the Contract, the Taking-Over Certificate shall be deemed to have been issued on the last day of that period.

條款 53

香港特別行政區土木工程合同一般條件：

「本工程」的完成

(1) 當「本工程」經已大體上完成，並令人滿意地通過任何在「合同」內所規定的最後測試時，「承建商」可以書面通知「工程師」此事，要求他就「本工程」頒發竣工證明書，並附帶在「保養期」內完成任何尚未完成的工程的承諾。「工程師」必須在收到通知書日期起計的21天內：

    (a) 頒發竣工證明書，說明「工程師」認為按照「合同」的規定已大體上完成「本工程」的日期，及由這證明書上所說明的竣工日期翌日起計「保養期」；或

    (b) 向「承建商」發出書面指令，說明「工程師」認為在頒發上述證明書前，「承建商」必須完成「工程師」需要他完成的所有工程，在此情況下，「承建商」不可再要求獲頒發竣工證明書，而本條第(2)款的條文將適用。

(2) 雖然有本條第(1)款的規定，當「工程師」認為「本工程」已大體上完成，並已令人滿意地通過任何在「合同」內規定的最後測試，「工程師」必須立刻頒發「本工程」的竣工證明書，及由這證明書所說明的竣工日期翌日起計「保養期」。

Commentary

This Clause stipulates the power and the obligation of the Engineer, and the procedure he shall follow to issue a certificate of completion when the Works or a Section is substantially completed or when a part of the Works is completed and is capable of being permanently used or occupied by the Employer. The certificate of completion has several important effects. Firstly, it brings the Contractor's liability to liquidated damages to an end. Secondly, it starts the Maintenance Period during which the Contractor is to complete the outstanding works and repair works under Clause 56. Thirdly, the risk of damage to the Works except the outstanding works now passes to the Employer under Clause 21(1). However, the Contractor's indemnity to third party damage under Clause 22, and the Contractor's liability concerning accident and injury to workmen under Clause 25 still continues. Regarding insurance coverage, both the Employer and the Contractor need to consider carefully their positions and take out any additional insurance where necessary.

Analysis and application

Clause 53(1) "When the Works have been substantially completed and have satisfactorily passed any final test that may be prescribed by the Contract, ..."

There is no common definition for substantial completion but in the context of engineering Works, readiness for "operational or functional occupation" by the Employer seems to be the basic test, since there are outstanding works remaining to be done.[1] For example, in a flyover project consisting also of some ground level roadworks, substantial completion will be achieved when the flyover is ready for opening to traffic, though there are still ground level roadworks to be carried out. However, if there are still very substantial ground level roadworks to be completed, the Engineer may consider certifying completion of the flyover only in order to better control the completion of the ground level roadworks.

Clause 53(1) "... accompanied by an undertaking to carry out any outstanding work during the Maintenance Period, ..."

Usually, the Contractor gives a brief list of outstanding works to be undertaken but it is submitted that such practice may lead to disputes when damage occurs to the outstanding works during the Maintenance Period. For example, if part of the surface protection to a slope is outstanding, and a major slip occurs, it is not sure whether the Contractor is still responsible for the major slip. Therefore, careful consideration by the Engineer is required in preparing the list of outstanding works and the extent of the Works to be defined as outstanding should preferably be referred to the Drawings.

Clause 53(1)(a) "issue a certificate of completion stating the date on which, in the Engineer's opinion, the Works were substantially completed ..."

It should be noted that the date of completion of the Works or Section is usually not the date of issue of the certificate of completion. Sub-clause (1) requires that the Works "have been substantially completed and have satisfactorily passed any final test" before the Contractor gives notice to request the Engineer for a certificate of completion. The date of completion, however, depends only on the Works being substantially completed, which could be much earlier than the date of passing the final test and the subsequent date of application by the Contractor and the issue of certificate by the Employer. The understanding of this concept is very important. Firstly, the application of liquidated damages depends on the date of completion. Secondly, the passing of responsibility for the care of Works under Clause 21(1) is related to the date of completion, but not the date of the issue of the certificate of completion under the HKSAR Clause. Thirdly, whether the Works are still continued to be insured may depend either on the date of completion or the date of the issue of the certificate of completion in accordance with the policy wording. Fourthly, the Contractor may postpone applying for the certificate of completion in order to recover any disruption and prolongation costs for which the Employer may be liable. However, such strategy may not be workable because under sub-clause (2), the Engineer is under a duty to issue a certificate of completion when the Works or Section have been substantially completed and satisfactorily passed any final test. Moreover, the Contractor has a general duty to mitigate his loss and he should obtain a certificate of completion as soon as possible and rearrange his resources to suit.[2]

Clause 53(1)(b) "give instructions in writing to the Contractor specifying all the work which, in the Engineer's opinion, is required to be done by the Contractor before such certificate can be issued ..."

This sub-clause applies when the Contractor applies for a certificate of completion prematurely. It should be noted that after the first application fails, the Contractor is not permitted to apply again but it would then be up to the Engineer who has a duty to issue the certificate when the requirements in sub-clauses 1(b) and (2) are satisfied.

Clause 53(2) This sub-clause has been explained in detail above.

One viaduct of this dual 3-lane road is completed. Can the Contractor apply for a partial completion certificate? What considerations should the Engineer bear in mind?

The HKSAR Conditions:

Completion of the Works

(continued)

(3) The Contractor shall carry out any outstanding work as soon as practicable after the issue of the certificate of completion or as reasonably directed by the Engineer and in any event before the expiry of the Maintenance Period. The Contractor's obligation to provide, service and maintain site offices, latrines and the like, shall continue for as long as may be required by the Engineer.

(4) The provisions of sub-clauses (1), (2) and (3) of this Clause shall apply equally to any Section.

(5) (a) The Engineer shall give a certificate of completion in respect of any part of the Works which has been completed to the satisfaction of the Engineer and is required by the Employer for permanent occupation or use before the completion of the Works or any Section.

　　(b) The Engineer, following a written request from the Contractor, may give a certificate of completion in respect of any substantial part of the Works which has been completed to the satisfaction of the Engineer before the completion of the Works or any Section and is capable of permanent occupation and/or permanent use by the Employer.

　　(c) When a certificate of completion is given in respect of a part of the Works such part shall be considered as completed and the Maintenance Period for such part shall commence on the day following the date of completion stated in such certificate.

(6) Any certificate of completion given in accordance with this Clause in respect of any Section or part of the Works shall not be deemed to certify completion of any ground or surface requiring reinstatement unless the certificate shall expressly so state.

(7) For the purposes of this Clause the term "Works" shall exclude any maintenance work executed in accordance with Clause 56.

Corresponding clauses in the ICE Conditions:

Clause 48

(3) If any substantial part of the Works has been occupied or used by the Employer other than as provided in the Contract the Contractor may request in writing and the Engineer shall issue a Certificate of Substantial Completion in respect thereof. Such certificate shall take effect from the date of delivery of the Contractor's request and upon the issue of such certificate the Contractor shall be deemed to have undertaken to complete any outstanding work in that part of the Works during the Defects Correction Period.

(4) (*Quoted on page 114*)

(5) A Certificate of Substantial Completion given in respect of any Section or part of the Works before completion of the whole shall not be deemed to certify completion of any ground or surfaces requiring reinstatement unless such certificate shall expressly so state.

Clause 49(1)

The undertaking to be given under Clause 48(1) may after agreement between the Engineer and the Contractor specify a time or times within which the outstanding work shall be completed as soon as practicable during the Defects Correction Period.

Corresponding clauses in the FIDIC Conditions:

Clause 10.2, paragraphs 1, 2 and 3

The Engineer may, at the sole discretion of the Employer, issue a Taking-Over Certificate for any part of the Permanent Works.

　　The Employer shall not use any part of the Works (other than as a temporary measure which is either specified in the Contract or agreed by both Parties) unless and until the Engineer has issued a Taking-Over Certificate for this part. However, if the Employer does use any part of the Works before the Taking-Over Certificate is issued:

(a)　the part which is used shall be deemed to have been taken over as from the date on which it is used,

(b)　the Contractor shall cease to be liable for the care of such part as from this date, when responsibility shall pass to the Employer, and

(c)　if requested by the Contractor, the Engineer shall issue a Taking-Over Certificate for this part.

After the Engineer has issued a Taking-Over Certificate for a part of the Works, the Contractor shall be given the earliest opportunity to take such steps as may be necessary to carry out any outstanding Tests on Completion. The Contractor shall carry out these Tests on Completion as soon as practicable before the expiry date of the relevant Defects Notification Period.

條款 53

香港特別行政區土木工程合同一般條件：

「本工程」的完成

（續）

(3)　「承建商」必須在頒發竣工證明書後，在實際可行的情況下盡快施行任何還未完成的工程，或按照「工程師」發給的合理指示進行這些工程，但無論如何，此等工程必須在「保養期」期滿前進行。只要「工程師」有所要求，「承建商」便有責任繼續提供、維修及維護「工地」的辦公室、洗手間及此類設施。

(4)　本條第(1)、(2)及(3)款同樣可用於「本工程」的任何「工段」。

(5)　(a)　如「本工程」的任何部分已令「工程師」滿意地完成，而「僱主」在「本工程」或其任何「工段」完成之前需要該部分作永久佔用或使用，「工程師」必須頒發該部分的竣工證明書。

　　　(b)　如在「本工程」或其任何「工段」完成之前，「本工程」的任何頗大部分已令「工程師」滿意地完成，及此部分可給「僱主」作永久佔用及/或永久使用，「工程師」在收到「承建商」的書面要求後，可頒發該部分的竣工證明書。

　　　(c)　當「工程師」頒發「本工程」某部分的竣工證明書後，這部分必須被視為完成，而這部分的「保養期」必須由上述證明書所說明的竣工日期翌日起計。

(6)　按照本條頒發的「本工程」任何「工段」或部分的任何竣工證明書，不可被視為核實已完成任何有需要恢復原狀的地面或表層的工程，但在證明書明文地說明的除外。

(7)　為本條的作用而言，「本工程」這詞不包括「承建商」根據第56條實施的任何保養工程。

Analysis and application

Clause 53(3)　"The Contractor shall carry out any outstanding work as soon as practicable after the issue of the certificate of completion or as reasonably directed by the Engineer and in any event before the expiry of the Maintenance Period …"

It is undesirable if the Contractor redeploys his resources to other contracts soon after the certificate of completion of the Works is issued, leaving a lot of outstanding works unattended which may attract public complaints and may even create safety problems. It is advisable that the Engineer demands a programme from the Contractor for the completion of the outstanding works because the Contractor is under a duty to "carry out any outstanding works as soon as practicable" and not to wait until the end of the Maintenance Period. As discussed before, if there are substantial works outstanding, the Engineer may issue a completion certificate for part of the Works which reduce the rate of potential liquidated damages in accordance with Clause 52, and the Contractor is still obliged to complete the remaining part of the Works in accordance with the date or extended date for completion.

Clause 53(5)(a) "The Engineer shall give a certificate of completion in respect of any part of the Works which has been completed … and is required by the Employer for permanent occupation or use …"

This sub-clause enables the Engineer to issue a certificate of completion when the Employer requires part of the completed Works for permanent occupation or use. The pre-requisite for the issue of such a partial completion certificate is: (1) the part of the Works being substantially completed and, (2) the part of the Works is required by the Employer for permanent occupation or use. The Engineer should note that once the part of the Works is occupied, the Employer will be responsible for the care of that part of the Works as "excepted risk" under Clause 21(4)(f).

Clause 53(5)(b) "The Engineer, following a written request from the Contractor, may give a certificate of completion in respect of any substantial part of the Works … completed to the satisfaction of the Engineer … and is capable of permanent occupation and/or permanent use by the Employer."

This sub-clause allows the Contractor to apply for a partial completion certificate. It should be noted that this applies only if a substantial part of the Works has been substantially completed and is capable of permanent occupation and/or permanent use by the Employer. The wording "substantial part of the Works" is important because it prevents the Contractor from making numerous applications for partial completion, for example, for completed road slabs once some part is completed in order that the Maintenance Period for those slabs could start early.

Clause 53(4), (6) and (7) These sub-clauses are self-explanatory.

Comparison with the ICE Conditions:

This ICE Clause 48 is very similar to the HKSAR Clause 53. The term "Certificate of Substantial Completion" is used in the former while the latter only refers to "certificate of completion" although both have the same meaning. One significant difference is that the ICE Clause does not obligate nor empower the Engineer to issue a Certificate of Substantial Completion to the Contractor when the Works are in his opinion substantially completed if the Contractor does not apply for the certificate. However, the ICE Clause empowers the Engineer to issue a Certificate of Substantial Completion in respect of part of the Works if he considers that part of the Works has been substantially completed and has passed any final test prescribed by the Contract. The ICE Clause 49(1) suggests that in the undertaking to complete the outstanding work, the Engineer and the Contractor shall agree on a time or times within which the outstanding work shall be completed.

Comparison with the FIDIC Conditions

This FIDIC Clause 10.1 is similar to the HKSAR Clause and the ICE Clause. However, the emphasis of this FIDIC Clause is on the procedures for taking over by the Employer. The Contractor can make an application for taking over not earlier than 14 days before the Works, in his opinion, will be completed. Then the Engineer has 28 days to consider the application to either issue the Taking-Over Certificate or reject the application, giving reasons and specifying the works to be done by the Contractor to enable the Taking-Over Certificate to be issued. It is submitted that this FIDIC procedure could minimize the period of grey area in which the responsibility for the care of the Works is indeterminate and possibly limit such period to 28 days. The FIDIC Clause 10.2 allows the Employer to take over part of the Permanent Works for use on the condition that the responsibility for the care of the Works passes to the Employer once the Works are taken into use and that the Taking-Over Certificate must be issued. This FIDIC Clause 10.2 has similar effect to the HKSAR Clause 53(5) and the ICE Clause 48(3), which also allow the Employer to occupy and use part of the Works which has been completed.

Notes

1. Ian Duncan Wallace, *A Commentary on the FIDIC International Standard Form of Engineering and Building Contract.* See also the commentary on Clause 48(2) of the ICE Conditions of Contract in Vivian Ramsey and Stephen Furst, *Keating on Building Contracts*, 6th Edition.
2. See the commentary on Clause 48 in Max Abrahamson, *Engineering Law and the ICE Contracts*, 4th Edition, p. 160.

The viaduct is completed and ready to be opened to traffic. However, the ground-level roads are still far from completion. Can the Engineer certify substantial completion of the whole of the Works? What are the Engineer's considerations?

Clause 54

Suspension of the Works

(1) The Contractor shall upon the written order of the Engineer suspend the progress of the Works or any part thereof for such time or times and in such manner as the Engineer may consider necessary and shall during such suspension properly protect and secure the Works so far as is necessary in the opinion of the Engineer.

(2) If upon written application by the Contractor to the Engineer, the Engineer is of the opinion that the Contractor has been involved in additional expenditure by reason of a suspension order given by the Engineer under this Clause then the Engineer shall ascertain the Cost incurred and shall certify in accordance with Clause 79, unless such suspension order is:

(a) otherwise provided for in the Contract, or

(b) necessary by reason of weather conditions affecting the safety or quality of the Works or any part thereof, or

(c) necessary by reason of some default on the part of the Contractor or any person carrying out the Works, or

(d) necessary for the proper execution of the Works or for the safety of the Works or any part thereof or for the safety and health of any person or the safety of any property on or adjacent to the Site in as much as such necessity does not arise from any act or default of the Engineer or the Employer or from any of the excepted risks defined in Clause 21.

Corresponding clauses in the ICE Conditions:

Clause 40

(1) The Contractor shall on the written order of the Engineer suspend the progress of the Works or any part thereof for such time or times and in such manner as the Engineer may consider necessary and shall during such suspension properly protect and secure the work so far as is necessary in the opinion of the Engineer. Except to the extent that such suspension is

(a) otherwise provided for in the Contract or

(b) necessary by reason of weather conditions or by some default on the part of the Contractor or

(c) necessary for the proper construction and completion or for the safety of the Works or any part thereof in as much as such necessity does not

arise from any act or default of the Engineer or the Employer or from any of the Excepted Risks defined in Clause 20(2)

then if compliance with the Engineer's instructions under this clause involves the Contractor in delay or extra cost the Engineer shall take such delay into account in determining any extension of time to which the Contractor is entitled under Clause 44 and the Contractor shall subject to Clause 53 be paid in accordance with Clause 60 the amount of such extra cost as may be reasonable. Profit shall be added thereto in respect of any additional permanent or temporary work.

Corresponding clauses in the FIDIC Conditions:

Clause 8.8

The Engineer may at any time instruct the Contractor to suspend progress of part or all of the Works. During such suspension, the Contractor shall protect, store and secure such part or the Works against any deterioration, loss or damage.

The Engineer may also notify the cause for the suspension. If and to the extent that the cause is notified and is the responsibility of the Contractor, the following Sub-Clause 8.9, 8.10 and 8.11 shall not apply.

Clause 8.9

If the Contractor suffers delay and/or incurs Cost from complying with the Engineer's instructions under Sub-

Clause 8.8 [*Suspension of Work*] and/or from resuming the work, the Contractor shall give notice to the Engineer and shall be entitled subject to Sub-Clause 20.1 [*Contractor's Claims*] to:

(a) an extension of time for any such delay, if completion is or will be delayed, under Sub-Clause 8.4 [*Extension of Time for Completion*], and

(b) payment of any such Cost, which shall be included in the Contract Price.

After receiving this notice, the Engineer shall proceed in accordance with Sub-Clause 3.5 [*Determinations*] to agree or determine these matters.

The Contractor shall not be entitled to an extension of time for, or to payment of the Cost incurred in, making good the consequences of the Contractor' failure to protect, store or secure in accordance with Sub-Clause 8.8 [*Suspension or Work*].

Clause 8.10
The Contractor shall be entitled to payment of the value (as at the date of suspension) of Plant and/or Materials which have not been delivered to Site, if:

(a) the work on Plant or delivery of Plant and/or Materials has been suspended for more than 28 days, and
(b) the Contractor has marked the Plant and/or Materials as the Employer's property in accordance with the Engineer's instructions.

條款 54

香港特別行政區土木工程合同一般條件：
「本工程」的擱置

(1) 「承建商」須按照「工程師」的書面命令，就「工程師」認為需要的一段或數段時間內及方式擱置「本工程」或其任何部分，及必須在此等工程擱置期間以「工程師」認為需要的方式妥善地保護及穩固「本工程」。

(2) 如「工程師」就「承建商」向他提出的書面申請，認為「承建商」由於「工程師」根據本條發出的工程擱置令涉及額外開支，「工程師」必須確實所引致的費用並根據第79條給予核實，除非該擱置命令是：

(a) 在「合同」內另有說明；或
(b) 由於影響「本工程」或其任何部分的安全或質量的天氣狀況而需要的；或
(c) 由於「承建商」或實施「本工程」的任何人的失誤而需要的；或
(d) 為妥善地實施「本工程」或為「本工程」或其任何部分的安全或在「工地」或「工地」毗鄰的任何人的健康或任何人或財產的安全而需要的，但此等需並非因為「工程師」或「僱主」的任何行為或失誤或因為第21條所訂定的豁免風險所引致的。

Commentary

Engineers are usually wary of ordering suspension of the Works or part thereof because such order will usually result in extension of time, payment of prolongation cost and other costs arising from idling of Plant, labour and disruption to the Works. Suspension may be ordered by the Engineer to cater for the situation which cannot be foreseen by the Engineer or the Employer, or have not been allowed for when drawing up the Contract. This may include some particular events which the Government of the HKSAR as the Employer may host, such as an international trade conference or a visit by high-profile political figures, necessitating the stopping of part or the whole of the construction work. However, when the suspension is ordered for reasons stated in items (a) to (d) of sub-clause 2, i.e., for safety or health reasons or due to the default of the Contractor, such as the unsuitable method of construction affecting the quality of permanent work, no extension of time nor cost compensation are allowed. Of course, there are situations which neither the Employer nor the Contractor are responsible, for example, villagers objecting to or preventing any further work. In such case, the Engineer has to be very careful about the course of action to be followed and he does not have a duty to order a suspension of the Works unless it is for safety or health reasons.

Analysis and application

Clause 54(1) "… suspend the progress of the Works or any part thereof for such time or times and in such manner as the Engineer may consider necessary …"

It is of course in the interest of the Employer, the Contractor and the project itself to suspend only the minimum extent of the Works for the shortest possible period and in a manner that would cause the least disruption to the progress of the Works or to the overall programme. Since any suspension has a major impact on the programme and on the financial aspect of the project, the Engineer must fully consult the Employer before ordering the suspension unless under emergency situations.

Clause 54(2)(d) "necessary for the proper execution of the Works or for the safety of the Works … in as much as such necessity does not arise from any act or default of the Engineer or the Employer or from any of the excepted risks defined in Clause 21."

Default of the Engineer may include the provision of wrong setting-out lines and levels affecting the proper execution of the Works, and the default of the Employer may include faulty design by his designer affecting any further construction or posing safety hazards. Excepted risks also include problematic design by the Engineer or default by the Engineer or the Employer.

Comparison with the ICE Conditions

This ICE Clause 40(1) is almost identical to the HKSAR Clause 54 except that the element of profit could be added to any extra expenses incurred by the Contractor.

Comparison with the FIDIC Conditions

These FIDIC Clauses are similar to the ICE Clause and the HKSAR Clause. However, the Engineer may notify the Contractor the cause of suspension and if the notified cause is the responsibility of the Contractor, time and cost compensation shall not be made. The FIDIC Clauses also explicitly provide for payment of the value of plant and materials which have not been delivered to the Site under the conditions in Clause 8.10(a) and (b).

 Q 28 The last row of soil nails has now been installed. However, a monitoring survey detects some movements of the slope. What should the Engineer consider?

 Q 29 The launching girder is employed for placing the precast deck units. Suppose it slides substantially but is stopped by the automatic locking device, causing slight injuries to two workmen. Subsequently the Labour Department inspects and suspends the Works. What course of action should the Engineer take?

Clause 55

Suspension Lasting More than 90 Days

If the progress of the Works or any part thereof is suspended on the written order of the Engineer and if written permission to resume work is not given by the Engineer within a period of 90 days after the date of suspension then the Contractor may, unless such suspension is occasioned by the circumstances described in Clause 54(2)(a) to (d), serve a notice in writing on the Engineer requiring permission within 28 days after the receipt of such notice to proceed with the Works or that part thereof in regard to which progress is suspended. If within the said 28 days the Engineer does not grant such permission the Contractor by a further notice in writing served on the Engineer may, but is not bound to, elect to treat the suspension where it affects part only of the Works as an omission of such part under Clause 60 or where it affects the Works as an abandonment of the Contract by the Employer.

Corresponding clauses in the ICE Conditions:

Clause 40

(2) If the progress of the Works or any part thereof is suspended on the written order of the Engineer and if permission to resume work is not given by the Engineer within a period of 3 months from the date of suspension then the Contractor may unless such suspension is otherwise provided for in the Contract or continues to be necessary by reason of some default on the part of the Contractor serve a written notice on the Engineer requiring permission within 28 days from the receipt of such notice to proceed with the Works or that part thereof in regard to which progress is suspended. If within the said 28 days the Engineer does not grant such permission the Contractor by a further written notice so served may (but is not bound to) elect to treat the suspension where it affects part only of the Works as an omission of such part under Clause 51 or where it affects the whole Works as an abandonment of the Contract by the Employer.

Corresponding clauses in the FIDIC Conditions:

Clause 8.11

If the suspension under Sub-Clause 8.8 [*Suspension of Work*] has continued for more than 84 days, the Contractor may request the Engineer's permission to proceed. If the Engineer does not give permission within 28 days after being requested to do so, the Contractor may, by giving notice to the Engineer, treat the suspension as an omission under Clause 13 [*Variations and Adjustments*] of the affected part of the Works. If the suspension affects the whole of the Works, the Contractor may give notice of termination under Sub-Clause 16.2 [*Termination by Contractor*].

條款 55

工程擱置超過90天

如「本工程」或其任何部分由於「工程師」的書面命令而擱置，而「工程師」並沒有在工程擱置的日期後90天內給予復工的書面允許，則除非此項工程擱置是由第54(2)(a)至(d)條所敘述的情況所引致，「承建商」可給予「工程師」書面通知，要求他在收到這通知後28天內允許恢復實施「本工程」或被擱置的部分。如在上述28天內「工程師」並無給予允許，「承建商」可以但並非必須再給予「工程師」另一書面通知，並作出如下選擇：當工程擱置影響「本工程」的部分時，將此部分根據第60條刪除；或當工程擱置影響到整個「本工程」時，可視為「僱主」放棄「合同」。

Commentary

This Clause simply states that if the suspension order is not lifted within 90 days, unless the cause of suspension is due to circumstances described in Clause 54(2)(a) to (d), the Contractor can give 28 days' notice to the Engineer requiring permission to proceed with the Works. If the Engineer does not do so, the Contractor can treat the suspension as an abandonment of the Contract by the Employer, or in the case that only part of the Works is affected, as an omission of that part of the Works.

Analysis and application

The interpretation of this Clause 55 is straightforward.

Comparison with the ICE Conditions

This ICE Clause 40(2) is, in essence, the same as the HKSAR Clause 55.

Comparison with the FIDIC Conditions

This FIDIC Clause 8.11 is similar to the HKSAR Clause 55 and the ICE Clause 40(2), but the time period which triggers notice to terminate is 84 days, instead of the 90 days under the HKSAR Conditions or 3 months under the ICE Conditions.

Clause 56

Execution of Work of Repair

(1) The Works shall at or as soon as practicable after the expiry of the Maintenance Period be delivered up to the Employer in the condition required by the Contract, fair wear and tear excepted.

(2) All maintenance work whether or not required urgently by the Engineer shall be carried out by the Contractor during the Maintenance Period or within 14 days after its expiry, and the Engineer may by notice in writing require the Contractor to carry out maintenance work including any work of repair or rectification, or make good any defect, imperfection, shrinkage, settlement or other fault identified within the Maintenance Period, and the Contractor shall carry out such work within the Maintenance Period or as soon as practicable thereafter and where the Engineer requires such maintenance work to be carried out urgently, the Contractor shall carry out such work in compliance with such terms contained in the notice imposed by the Engineer as the Engineer may consider necessary and reasonable in the circumstances.

(3) All such work shall be carried out by the Contractor at his own expense if the necessity for such work shall, in the Engineer's opinion, be due to the use of materials or workmanship not in accordance with the Contract or due to neglect or failure on the part of the Contractor to comply with any obligation expressed or implied on the Contractor's part under the Contract. If in the opinion of the Engineer such necessity shall be due to any other cause, the Engineer shall value the work as if it were a variation ordered in accordance with Clause 60, and shall certify in accordance with Clause 79.

Corresponding clauses in the ICE Conditions:

Clause 49

(2) The Contractor shall deliver up to the Employer the Works and each Section and part thereof at or as soon as practicable after the end of the relevant Defects Correction Period in the condition required by the Contract (fair wear and tear excepted) to the satisfaction of the Engineer. To this end the Contractor shall as soon as practicable carry out all work of repair amendment reconstruction rectification and making good of defects of whatever nature as may be required of him in writing by the Engineer during the relevant Defects Correction Period or within 14 days after its expiry as a result of an inspection made by or on behalf of the Engineer prior to its expiry.

(3) All work required under sub-clause (2) of this Clause shall be carried out by the Contractor at his own expense if in the Engineer's opinion it is necessary due to the use of materials or workmanship not in accordance with the Contract or to neglect or failure by the Contractor to comply with any of his obligations under the Contract. In any other event the value of such work shall be ascertained and paid for as if it were additional work.

Corresponding clauses in the FIDIC Conditions:

Clause 11.1

In order that the Works and Contractor's Documents, and each Section, shall be in the condition required by the Contract (fair wear and tear excepted) by the expiry date of the relevant Defects Notification Period or as soon as practicable thereafter, the Contractor shall:

(a) complete any work which is outstanding on the date stated in a Taking-Over Certificate, within such reasonable time as is instructed by the Engineer, and

(b) execute all work required to remedy defects or damage, as may be notified by (or on behalf of) the Employer on or before the expiry date of the Defects Notification Period for the Works or Section (as the case may be).

If a defect appears or damage occurs, the Contractor shall be notified accordingly, by (or on behalf of) the Employer.

Clause 11.2

All work referred to in sub-paragraph (b) of Sub-Clause 11.1 [*Completion of Outstanding Work and Remedying Defects*] shall be executed at the risk and cost of the Contractor, if and to the extent that the work is attributable to:

(a) any design for which the Contractor is responsible,

(b) Plant, Materials or workmanship not being in accordance with the Contract, or

(c) failure by the Contractor to comply with any other obligation.

If and to the extent that such work is attributable to any other cause, the Contractor shall be notified promptly by (or on behalf of) the Employer, and Sub-Clause 13.3 [*Variation Procedure*] shall apply.

條款 56

香港特別行政區土木工程合同一般條件：
修理工程的實施

(1) 「本工程」必須在「保養期」期滿後，或在「保養期」期滿後在實際可行的情況下盡快按照「合同」所規定的狀況交給「僱主」，但正常損耗則除外。

(2) 無論「工程師」有否要求緊急進行，一切保養工程必須由「承建商」在「保養期」內或在「保養期」期滿後14天內進行。「工程師」可以書面通知，要求「承建商」施行保養工程，包括任何修補或更正工程，或修復在「保養期」內認定的任何缺陷、欠完善之處、收縮、沉降或其他錯誤，「承建商」並必須在「保養期」內或在「保養期」期滿後在實際可行的情況下盡快進行這些工程；如「工程師」要求緊急施行這些保養工程，「承建商」必須在施行這些工程時遵照該項通知所載述「工程師」認為在當時情況下是合理地必要加入的條款。

(3) 如「工程師」認為所有此等工程是由於「承建商」使用不符合「合同」所規定的材料或技術水平，或由於「承建商」本身的疏忽或沒有遵照「合同」明示或隱含在任何「承建商」的責任而必須進行的，則所有此等工程必須全部由「承建商」自費施行。如「工程師」認為此等需要是由於任何其他原因所引致，「工程師」必須釐定其價值就如此等工程是根據第60條所發出的變更命令無異，並必須根據第79條加以核實。

Commentary

There are many misconceptions about the Maintenance Period and many engineers equate the Maintenance Period to a period in which the Contractor is still responsible for the defects arising from the Works. This is understandable because the Government's form of Conditions of Contract has at one time changed the term "Maintenance Period" to "Defects Liability Period." In reality, during the Maintenance Period, the Contractor has a right (and obligation) to repair his own defects in the Works, instead of another contractor being called upon to carry out the repair works, possibly at a higher cost with the Employer subsequently charging the Contractor for such repair works. After the Maintenance Period, if defects in the Works are discovered which is due to the Contractor's poor workmanship, such right of the Contractor expires and the Employer is then at liberty to employ his own contractor to carry out the repair and charge the Contractor for such repair. The Contractor's liability for defective works depends on the *Limitation Ordinance* and lasts for 6 years and 12 years after completion of the Works for simple contract and contract under seal respectively. On the other hand, if the defects are due to some other causes which the Contractor is not responsible, the repair works are still to be carried out by the Contractor during the Maintenance Period but the Employer will then be responsible for the Cost as if the repair works were a variation ordered under Clause 60. Apart from rectifying defects, the Employer may want some enhancement of the Works and this will be a variation. The present Clause does not appear to have conferred a right to the Engineer to order variations during the Maintenance Period but the HKSAR contracts now incorporate a Special Condition which allows the ordering of variations during the Maintenance Period.

Analysis and application

Clause 56(1) "The Works shall at … the expiry of the Maintenance Period be delivered up to the Employer in the condition required by the Contract, fair wear and tear excepted."

The basic obligation to complete the Works is satisfied once the Engineer certifies the completion of the Works. The sub-clause applies to outstanding works and defects, which are discovered during the Maintenance Period. The wording "in the condition required by the Contract" makes it clear that even if defects exist but have not been discovered by the Engineer or Employer when completion is certified and the Works taken over, the Contractor is still responsible for any non-compliance of the Contract.

Clause 56(2) "All maintenance work whether or not required urgently by the Engineer shall be carried out by the Contractor during the Maintenance Period or within 14 days after its expiry, …"

This first part of the sub-clause sets out the time limit for the Contractor to complete all maintenance work, whether urgent or not.

Clause 56(2) "… and the Engineer may by notice in writing require the Contractor to carry out maintenance work including any work of repair or rectification or make good any defect, imperfection, shrinkage, settlement or other fault identified within the Maintenance Period, …"

The wording appears to suggest a definition for maintenance work, which include any work of repair, rectification of defects, imperfection, shrinkage, settlement or other fault identified. Although these are not exclusive of other works, it is submitted that there appears to be no other works which fall into the definition of maintenance work for a pure civil engineering contract. For a contract with E & M components or for E & M contracts, regular maintenance of the Plant may also form part of the maintenance work.

Clause 56(2) "… where the Engineer requires such maintenance work to be carried out urgently, the Contractor shall carry out such work in compliance with such terms contained in the notice imposed by the Engineer as the Engineer may consider necessary and reasonable in the circumstances."

This part of the sub-clause enables the Engineer to require the Contractor to carry out maintenance in the manner and timing specified by the Engineer, subject to being necessary and reasonable in the circumstances.

Clause 56(3) This sub-clause is self-explanatory and simply states that the cost of repair works shall be borne by the Contractor if due to his own fault, otherwise the cost will be borne by the Employer.

 The main tunnel Works are completed and the Engineer has issued a certificate of completion. However, the Transport Department wants to install a new automatic smoke detection system. Can the Engineer order such work? How can he ensure that the Works are completed within the Maintenance Period?

 31 Four kilometres of twin bridge decks have been completed but two short sections of parapets are still outstanding pending rectification of lapping bars. Can the Engineer certify completion and leave the parapets as outstanding works to be completed in the Maintenance Period?

(4) If the Contractor fails to carry out any outstanding work as required by Clause 53(3) or fails to carry out any maintenance work and in such terms (if any) as required by the Engineer under sub-clause (2) of this Clause the Employer shall be entitled after giving reasonable notice in writing to the Contractor, to have such work carried out by his own workers or by other contractors and if such work is work which the Contractor would have been required to carry out at his own expense the Employer shall be entitled to recover from the Contractor the expenditure incurred in connection therewith.

Corresponding clauses in the ICE Conditions:

Clause 49

(4) If the Contractor fails to do any such work as aforesaid the Employer shall be entitled to carry out that work by his own workpeople or by other contractors and if it is work which the Contractor should have carried out at his own expense the Employer shall be entitled to recover the cost thereof from the Contractor and may deduct the same from any monies that are or may become due to the Contractor.

Corresponding clauses in the FIDIC Conditions:

Clause 11.4

If the Contractor fails to remedy any defect or damage within a reasonable time, a date may be fixed by (or on behalf of) the Employer, on or by which the defect or damage is to be remedied. The Contractor shall be given reasonable notice of this date.

If the Contractor fails to remedy the defect or damage by this notified date and this remedial work was to be executed at the cost of the Contractor under Sub-Clause 11.2 [*Cost of Remedying Defects*], the Employer may (at his option):

(a) carry out the work himself or by others, in a reasonable manner and at the Contractor's cost, but the Contractor shall have no responsibility for this work; and the Contractor shall subject to Sub-Clause 2.5 [*Employer's Claims*] pay to the Employer the costs reasonably incurred by the Employer in remedying the defect or damage;

(b) require the Engineer to agree or determine a reasonable reduction in the Contract Price in accordance with Sub-Clause 3.5 [*Determinations*]; or

(c) if the defect or damage deprives the Employer of substantially the whole benefit of the Works or any major part of the Works, terminate the Contract as a whole, or in respect of such major part which cannot be put to the intended use. Without prejudice to any other rights, under the Contract or otherwise, the Employer shall then be entitled to recover all sums paid for the Works or for such part (as the case may be), plus financing costs and the cost of dismantling the same, clearing the Site and returning Plant and Materials to the Contractor.

條款 56
修理工程的實施

（續）

(4) 如「承建商」未能按照第53條第(3)款的規定施行任何未完成的工程，或未能按照本條第(2)款及未能按照「工程師」根據本條第(2)款訂定的條件（如有的話）施行任何保養工程，則「僱主」有權以書面向「承建商」發出合理時間的通知後，安排由自己的工人或其他承建商施行這些工程，及如這些工程本應必須由「承建商」自費施行，「僱主」便有權向「承建商」追討由於與此等工程有關的事宜所引致的開支。

Analysis and application

Clause 56(4) This sub-clause is also self-explanatory and simply states that if the Contractor fails to carry out any outstanding works or maintenance works and in terms as required by the Engineer, such as following a time table, the Employer can, after giving reasonable notice in writing, arrange his own workers to carry out the said works and recover money from the Contractor.

Comparison with the ICE Conditions

This ICE Clause is very similar to the HKSAR Clause. However, it does not have a special provision for urgent maintenance works for which the Engineer can impose terms, presumably including the timing for the carrying out of such works as in the case of the HKSAR Clause 56(2). The term "Defects Correction Period" is used which means the defects, whatever the cause, are to be carried out by the Contractor and it is submitted that this term is more appropriate.

Comparison with the FIDIC Conditions

This FIDIC Clause adopts the term "Defects Notification Period" because notice of defects is administratively very significant during the period. The Contractor has to complete all outstanding works and all works required to remedy defects or damage. The Employer has a duty to notify the Contractor if a defect appears or damage occurs (before the expiry date of the Defects Notification Period)] and then after being notified, the Contractor shall be obligated to carry out the repair works. Like the HKSAR Clause and the ICE Clause, the FIDIC Clause also delineates the responsibility for the cost of the repair works, depending on whom the faults belong to. The FIDIC Clause 11.4 describes in detail on the procedures to follow if the Contractor fails to remedy any defects or damage after being given reasonable notice. The Employer may carry out the repair works himself and recover the cost from the Contractor, or require the Engineer to agree or determine a reduction in the Contract Price or if the defects or damage are fatal, terminate the Contract as a whole. Such detailed procedures are absent in both the HKSAR Conditions or the ICE Conditions.

Investigating Defects

(1) At any time prior to the issue of the maintenance certificate in accordance with Clause 80 the Contractor shall, if instructed by the Engineer in writing, investigate the cause of any defect, imperfection or fault under the directions of the Engineer.

   Provided that if the Engineer at his absolute discretion so decides, the Employer shall be entitled, after giving reasonable notice in writing to the Contractor, to have such investigation carried out by his own workers or by other contractors.

(2) If such defect, imperfection or fault shall be one for which the Contractor is liable in accordance with the provisions of the Contract, the expense incurred in investigating as aforesaid shall be borne by the Contractor and he shall in such case repair, rectify and make good such defect, imperfection or fault together with any consequential damage at his own expense.

(3) If such defect, imperfection or fault shall be one for which the Contractor is not so liable, then the Engineer shall value any investigation and remedial work carried out by the Contractor as aforesaid in accordance with Clause 61, and shall certify in accordance with Clause 79.

Corresponding clauses in the ICE Conditions:

Clause 50

The Contractor shall if required by the Engineer in writing carry out such searches tests or trials as may be necessary to determine the cause of any defect imperfection or fault under the directions of the Engineer. Unless the defect imperfection or fault is one for which the Contractor is liable under the Contract the cost of the work carried out by the Contractor as aforesaid shall be borne by the Employer. If the defect imperfection or fault is one for which the Contractor is liable the cost of the work carried out as aforesaid shall be borne by the Contractor and he shall in such case repair rectify and make good such defect imperfection or fault at his own expense in accordance with Clause 49.

Corresponding clauses in the FIDIC Conditions:

Clause 11.8

The Contractor shall, if required by the Engineer, search for the cause of any defect, under the direction of the Engineer. Unless the defect is to be remedied at the cost of the Contractor under Sub-Clause 11.2 [*Cost of Remedying Defects*], the Cost of the search plus reasonable profit shall be agreed or determined by the Engineer in accordance with Sub-Clause 3.5 [*Determinations*] and shall be included in the Contract Price.

條款 58	香港特別行政區土木工程合同一般條件：

調查缺陷

(1) 在「工程師」根據第80條頒發保養證明書前的任何時間，如「工程師」以書面發出指令，「承建商」必須在「工程師」指引下調查任何缺陷，欠完善之處、或錯誤的原因。

　　但如「工程師」行使他的絕對酌情權而如此決定，則「僱主」有權以書面向「承建商」發出合理時間的通知後，安排由自己的工人或其他承建商進行這些調查。

(2) 如這些缺陷、欠完善之處或錯誤本應必須由「承建商」按照「合同」條文負責，則這些調查所引致的開支須由「承建商」負擔，「承建商」並須在此情況下自費修補、更正及修復此等缺陷、欠完善之處或錯誤連同任何因而導致的損毀。

(3) 如此等缺陷、欠完善之處或錯誤本應無須由「承建商」負責任，「工程師」必須根據第61條對「承建商」所進行的此等調查及補救工程釐定價值，並必須根據第79條加以核實。

Commentary

This Clause empowers the Engineer to require the Contractor to investigate the cause of any defects or faults in the Works under the direction of the Engineer. This is despite the fact that the Site has already been handed back to the Employer. Such power lasts until the issue of the maintenance certificate which could be many months or sometimes even years after the expiry of the Maintenance Period, because the maintenance certificate can only be issued after all outstanding works and defects are rectified. As to the Cost involved, the Contractor shall bear such Cost if it is found that the defects are due to bad workmanship or fault of the Contractor, otherwise the Cost shall be borne by the Employer.

Analysis and application

Clause 58(1) "At any time prior to the issue of the maintenance certificate … the Contractor shall, … investigate the cause of any defect, … under the direction of the Engineer …"

As explained in the commentary, the issue of the maintenance certificate may be a couple of years after the completion of the Works. This is a very useful clause for the Employer if after a long time it is subsequently found that part of the completed Works is defective but the cause is not known, the Contractor could be called back to carry out the investigation work under the direction of the Engineer. For example, if a major filled slope is found to be subsiding a couple of years after completion before the maintenance certificate is issued, the Employer can rely on this Clause provided that the Engineer is still around to give instructions and directions.

Clause 58(1) The proviso to this sub-clause is straightforward, giving the Employer the option to use his own workers or other contractors to carry out the investigation.

Clause 58(2) and (3) These two sub-clauses are also straightforward, delineating the liability for the Cost of investigation between the Contractor and the Employer.

Comparison with the ICE Conditions

This ICE Clause does not mention about the time when the power is exercisable but probably it applies to the Maintenance Period only and not up to the issue of the maintenance certificate as in the case of the HKSAR Clause. Otherwise the ICE Clause is similar to the HKSAR Clause.

Comparison with the FIDIC Conditions

This FIDIC Clause is very similar to the ICE Clause. One difference is that reasonable profit can be included in the Cost of the search if it is subsequently found that the Contractor is not liable for such defects.

Clause 59

The HKSAR Conditions:

Bills of Quantities and Measurement

(1) Except where any statement in the Bills of Quantities expressly shows to the contrary the Bills of Quantities shall be deemed to have been prepared and measurements shall be made according to the procedures set forth in the Method of Measurement stated in the Preamble to the Bills of Quantities.

(2) The quantities set out in the Bills of Quantities are estimated quantities and they are not to be taken as the actual and correct quantities of the work to be executed.

(3) Any error in description in the Bills of Quantities or item omitted therefrom shall not vitiate the Contract nor release the Contractor from the execution of the whole or any part of the Works according to the Drawings and Specification or from any of his obligations or liabilities under the Contract. The Engineer shall correct any such error or omission, shall ascertain the value of the work actually carried out in accordance with Clause 61, and shall certify in accordance with Clause 79.

Provided that there shall be no rectification of any error, omission or wrong estimate in any description, quantity or rate inserted by the Contractor in the Bills of Quantities.

Corresponding clauses in the ICE Conditions:

Clause 57

Unless otherwise provided in the Contract or unless general or detailed description of the work in the Bill of Quantities or any other statement clearly shows to the contrary the Bill of Quantities shall be deemed to have been prepared and measurements shall be made according to the Procedure set out in the "Civil Engineering Standard Method of Measurement Third Edition 1991" approved by the Institution of Civil Engineers and the Federation of Civil Engineering Contractors in association with the Association of Consulting Engineers or such later or amended edition thereof as may be stated in the Appendix to the Form of Tender to have been adopted in its preparation.

Clause 55

(1) The quantities set out in the Bill of Quantities are the estimated quantities of the work but they are not be taken as the actual and correct quantities of the Works to be carried out by the Contractor in fulfillment of his obligations under the Contract.

(2) No error in description in the Bill of Quantities or omission therefrom shall vitiate the Contract nor release the Contractor from the carrying out of the whole or any part of the Works according to the Drawings and Specification or from any of his obligations or liabilities under the Contract. Any such error or omission shall be corrected by the Engineer and the value of the work actually carried out shall be ascertained in accordance with Clause 52(2) or (3). Provided that there shall be no rectification of any errors omissions or wrong estimates in the descriptions rates and prices inserted by the Contractor in the Bill of Quantities.

Corresponding clauses in the FIDIC Conditions:

Clause 12.2

Except as otherwise stated in the Contract and notwithstanding local practice:

(a) measurement shall be made of the net actual quantity of each item of the Permanent Works, and

(b) the method of measurement shall be in accordance with the Bill of Quantities or other applicable Schedules.

條款 59	*香港特別行政區土木工程合同一般條件：* 「工程量清單」及計量

(1) 除「工程量清單」內的任何敘述明示相反意思，「工程量清單」必須被視為已按照其序言中的計量方法所列的程序擬備，及計量亦必須按照此等程序進行。

(2) 「工程量清單」所列出的數量是估計數量，不可被視為所要實施的工程的真實及正確數量。

(3) 「工程量清單」內任何敘述的錯誤或項目的遺漏皆不會令「合同」無效，也不會解除「承建商」必須按照「圖則」及「規格」實施「本工程」的全部或部分的責任，也不會解除「承建商」在「合同」上的任何責任或法律責任。「工程師」必須改正任何此等錯誤或遺漏，必須根據第61條確定已實際施行的工程價值，並必須根據第79條加以核實。

但是，在「工程量清單」內由「承建商」加入的任何敘述、數量或價率如有任何錯誤、遺漏或估計錯誤，皆不可糾正。

Commentary

This Clause 59, especially sub-clauses (1) to (4), sets out in detail the principles, which form the basis of the re-measurement contract to which this Condition of Contract applies. Firstly, the Preamble to the Bill of Quantities must make reference to the Method of Measurement, which itself is a very substantial document. The requirement that the Bill of Quantities are prepared and measurements made in accordance with the procedures set forth in the Method of Measurement gives rise to a lot of claims and disputes, the subject of many arbitrations and Court cases. Secondly, the quantities are expressly stated not to be taken as the actual and correct quantities. However, if there is any error in the description of the Bill of Quantities or any items omitted from it in relation to the Works which are described in the Drawings and/or the Specification, the Contractor is entitled to any losses arising from making corrections by the Engineer. The Engineer must make a valuation of the related works in accordance with Clause 61. This also gives rise to many major disputes. Thirdly, the quantities in the Bill of Quantities are expressly stated not to be taken as actual and correct quantities and the Final Contract Sum is calculated by reference to the actual quantities of work completed based on the rates set out in the Bill of Quantities. The majority of these rates are priced by the Contractor during the Tender and there are also other rates determined generally in accordance with sub-clauses (3), 4(b) and Clause 61. Fourthly, if the actual quantities executed deviates substantially from those given in the Bill of Quantities, and the increase or decrease "itself" render the rate for the item unreasonable, the Contractor is entitled to an adjustment of the rate.

Analysis and application

Clause 59(1) "Except where any statement in the Bills of Quantities shows to the contrary …"

Though the Method of Measurement has been incorporated into the Contract by reference to the Preamble to the Bill of Quantities, its procedures and provisions can be overridden by either the Particular Preamble to the Bill of Quantities specifically written for the Contract or any statement in the Bill of Quantities item itself. The wording "shows to the contrary" gives a wider scope for interpretation of the intention of the drafter of the Bill of Quantities than if the wording "stated to the contrary" is used.[1] Therefore, if the description of any item in the Bill of Quantities is not ambiguous and quite clearly shows the intention about the coverage of any item and/or the method of measurement, the Contractor may not rely on any contrary provisions in the Method of Measurement to claim extra costs provided that he has not been misled as a matter of fact.

Clause 59(2) This sub-clause is straightforward, but see also the commentary to sub-clause 4(b).

Clause 59(3) "Any error in the description in the Bills of Quantities or item omitted therefrom shall not vitiate the Contract nor release the Contractor from execution … of the Works according to the Drawings and Specification …"

It is very important for those responsible for the preparation of the Bill of Quantities to thoroughly understand the Method of Measurement, whether its general principle, the general preamble, the itemization of the Works and the item coverage, because any incompatibility of this Bill of Quantities with the Method of Measurement may be the subject of a claim or potential dispute. For example, if a certain kind of test for piling work is required in the Particular Specification, which, however, has not been included in the item coverage of the piling work nor has the test been included as a separate item, the Contractor may be entitled to compensation arising from a missing item. In the past, there have been many claims arising from missing item as a result of provision for sinking boreholes for establishing founding levels of bored piles not being separately itemized. Likewise, if the Method of Measurement requires different stages or kinds of work to be separately itemized, for example, mobilization, moving and removal of certain piling equipment for different types of piles, any attempt to combine the operations for the of piles without a clear statement in the Particular Preamble or Bill of Quantities could attract claims. The valuation of the claims would be made by the Engineer in accordance with Clause 61.

Clause 59 "Provided that there shall be no rectification of any error, omission or wrong estimate in any description, quantity or rate inserted by the Contractor in the Bills of Quantities."

Usually, the Contractor is only required to insert the rates in the Bill of Quantities during the tender and any error or omission is corrected in accordance with the Condition of Tender before the Contract is signed. Therefore, this proviso may have only very limited application.

 32 How is the central median of the dual-three lanes flyover measured? Do you need to write any Particular Preamble?

 A median drainage to collect water from the wide bridge deck is unfortunately missed out from the Drawing. It is described in the Specification and an item has been allowed in the Bill of Quantities. Who is responsible for the delay and additional expenses of the rectification work?

Bills of Quantities and Measurement

(4) (a) For the purpose of calculating the Final Contract Sum the Engineer shall ascertain and determine by measurement the quantity of work executed in accordance with the Contract. Subject to (b) of this sub-clause such work shall be valued at the rates set out in the Bills of Quantities or if there are no appropriate rates in the Bills of Quantities then at other rates determined in accordance with the Contract.

(b) Should the actual quantity of work executed in respect of any item be substantially greater or less than that stated in the Bills of Quantities (other than an item included in the daywork schedule if any) and if in the opinion of the Engineer such increase or decrease of itself shall render the rate for such item unreasonable or inapplicable, the Engineer shall determine an appropriate increase or decrease of the rate for the item using the Bills of Quantities rate as the basis for such determination and shall notify the Contractor accordingly.

Corresponding clauses in the ICE Conditions:

Clause 56

(1) The Engineer shall except as otherwise stated ascertain and determine by measurement the value in accordance with the Contract of the work done in accordance with the Contract.

(2) Should the actual quantities carried out in respect of any item be greater or less than those stated in the Bill of Quantities and if in the opinion of the Engineer such increase or decrease of itself shall so warrant the Engineer shall after consultation with the Contractor determine an appropriate increase or decrease of any rates or prices rendered unreasonable or inapplicable in consequence thereof and shall notify the Contractor accordingly.

Corresponding clauses in the FIDIC Conditions:

Clause 12.3

Except as otherwise stated in the Contract, the Engineer shall proceed in accordance with Sub-Clause 3.5 [*Determinations*] to agree or determine the Contract Price by evaluating each item of work, applying the measurement agreed or determined in accordance with the above Sub-Clause 12.1 and 12.2 and the appropriate rate or price for the item.

For each item of work, the appropriate rate or price for the item shall be the rate or price specified for such item in the Contract or, if there is no such item, specified for similar work. However, a new rate or price shall be appropriate for an item or work if:

(a) (i) the measured quantity of the item is changed by more than 10% from the quantity of this item in the Bill of Quantities or other Schedule,
(ii) this change in quantity multiplied by such specified rate for this item exceeds 0.01% of the Accepted Contract Amount,
(iii) this change in quantity directly changes the Cost per unit quantity of this item by more than 1%, and
(iv) this item is not specified in the Contract as a "fixed rate item";

or

(b) (i) the work is instructed under Clause 13 [*Variations and Adjustments*],
(ii) no rate or price is specified in the Contract for this item, and
(iii) no specified rate or price is appropriate because the item of work is not of similar character, or is not executed under similar conditions, as any item in the Contract.

Each new rate or price shall be derived from any relevant rates or prices in the Contract, with reasonable adjustments to take account of the matters described in sub-paragraph (a) and/or (b), as applicable. If no rates or prices are relevant for the derivation of new rate or price, it shall be derived from the reasonable Cost of executing the work, together with reasonable profit, taking account of any other relevant matters.

Until such time as an appropriate rate or price is agreed or determined, the Engineer shall determine a provisional rate or price for the purposes of Interim Payment Certificates.

香港特別行政區土木工程合同一般條件：

「工程量清單」及計量

(4) (a) 計算「最終合同金額」，「工程師」必須用計量方式確定及釐定根據「合同」實施的工程數量。除本款(b)段別有規定外，此等工程必須以「工程量清單」內列出的價率釐定價值；如「工程量清單」內並無適用的價率，則按照「合同」而定出的其他價率釐定價值。

(b) 如某項目已實施的工程的實際數量有頗大幅度多於或少於「工程量清單」內所列出的數量（日工表上如包括此項目除外），而「工程師」認為此等數量的增加或減少本身會導致此項目的價率變成不合理或不適用，則「工程師」必須按照「工程量清單」內的價率釐定其適當的增加或減少，並就此通知「承建商」。

Analysis and application

Clause 59(4)(a) "… or if there are no appropriate rates in the Bills of Quantities then at other rates determined in accordance with the Contract."

As described in the commentary, there are generally three situations under which a new rate other than that given in the Bill of Quantities shall be fixed by the Engineer. The first situation arises from any missing item or error in the description in the Bill of Quantities. The means of rectification is given in Clause 59(5)(a) and (5)(b) which the Engineer is obliged to value under Clause 61. The second situation is for any substantial deviation of actual quantities from those shown in the Bill of Quantities and re-valuation of the rate is provided for under Clause 59(4)(b) described later in the commentary to this Clause. The third situation is the valuation of works ordered under Clause 60, which shall also be made by the Engineer in accordance with Clause 61.

Clause 59(4)(b) "Should the actual quantity of work … be substantially greater or less than that stated in the Bills of Quantities … and if in the opinion of the Engineer such increase or decrease of itself shall render the rate for such item unreasonable or inapplicable, the Engineer shall determine an appropriate increase or decrease of the rate for the item using the Bills of Quantities rate as the basis …"

This sub-clause is one of the most interesting and yet most contentious provision in the re-measurement Contract because it allows the rate priced by the Contractor in his Tender to be subsequently altered if the actual quantities of work deviate substantially from those given in the Bill of Quantities. Many contractors have a pricing strategy for re-measurement contract in that if they anticipate any underestimate of quantities, they would put high rates and vice versa and this enables them to submit a competitive tender to win the job and recoup sometimes tens of millions of dollars through substantial increase or decrease of quantities. The question therefore arises as to how this sub-clause operates if a rate entered by the Contractor is several times or even hundreds of times more expensive than the market rate or vice versa. To answer the question, we look at two conditions, which must be satisfied before this sub-clause can operate. The first condition is that there must be a substantial increase or decrease. The FIDIC Conditions specify among other criteria that a change of more than 10% of the quantities to be regarded as substantial but there is no defined limit for the HKSAR Conditions. The fact that "substantiality" is not defined imparts some flexibility for the Engineer to decide whether any given change in quantity fulfils this condition for rate adjustment. A very large percentage increase in low value item might not be substantial while a smaller percentage increase in a high value item might be. A recent arbitration considers that a change of over 40% in the quantity in site clearance to be substantial. The second condition is that the substantial change in quantity must "of itself" render the rate for the item unreasonable or inapplicable. It is considered that an increase in quantity imposing hardship on the Contractor who originally priced the item too low for any reason, or giving rise to large profit to the Contractor who priced the item very high does not warrant any change in rate because it is not the change of quantity "of itself" render the rate unreasonable or inapplicable.[2] To satisfy the second condition, the change in quantity must result in either a change of method of working or the economics of a working method.[3] An example is a decrease in quantity resulting in machine excavation being changed to hand excavation or the Cost of mobilization of the excavator being rendered less economical than originally envisaged. In a Court decision which was upheld in the Court of Appeal, it was stated that the test for the application of the ICE Clause 56(2) (almost identical to the HKSAR Clause 59(4)(b)) is whether the "increase or decrease of itself" shall so warrant, and not an error built into the rate or price that makes it reasonable to consider whether the rate or price should be altered.[4]

The HKSAR Conditions:

Bills of Quantities and Measurement

(5) (a) When any part of the Works is required to be measured the Engineer shall inform the Contractor who shall forthwith attend or send a representative to assist the Engineer in making such measurement and shall furnish all particulars required by him. Should the Contractor not attend or neglect or omit to send such representative then the measurement made by the Engineer shall be taken to be the correct measurement of the work.

(b) For the purpose of measuring such permanent work as is to be measured by records and Drawings the Engineer's Representative shall prepare records and Drawings month by month of such work and the Contractor, as and when called upon to do so in writing, shall within 14 days attend to examine and agree such records and Drawings with the Engineer's Representative and shall sign the same when so agreed and if the Contractor does not so attend to examine and agree any such records and Drawings they shall be taken to be correct.

(c) If after examination of such records and Drawings the Contractor does not agree the same or does not sign the same as agreed they shall nevertheless be taken to be correct unless the Contractor shall, within 14 days of such examination, lodge with the Engineer for a decision by the Engineer a statement in writing of the respects in which such records and Drawings are claimed by the Contractor to be incorrect.

Corresponding clauses in the ICE Conditions:

Clause 56

(3) The Engineer shall when he requires any part or parts of the work to be measured give reasonable notice to the Contractor who shall attend or send a qualified agent to assist the Engineer or the Engineer's Representative in making such measurement and shall furnish all particulars required by either of them. Should the Contractor not attend or neglect or omit to send such agent then the measurement made by the Engineer or approved by him shall be taken to be the correct measurement of the work.

Corresponding clauses in the FIDIC Conditions:

Clause 12.1

The Works shall be measured, and valued for payment, in accordance with this Clause.

Whenever the Engineer requires any part of the Works to be measured, reasonable notice shall be given to the Contractor's Representative, who shall:

(a) promptly either attend or send another qualified representative to assist the Engineer in making the measurement, and

(b) supply any particulars requested by the Engineer.

If the Contractor fails to attend or send a representative, the measurement made by (or on behalf of) the Engineer shall be accepted as accurate.

Except as otherwise stated in the Contract, wherever any Permanent Works are to be measured from records, these shall be prepared by the Engineer. The Contractor shall, as and when requested, attend to examine and agree the records with the Engineer, and shall sign the same when agreed. If the Contractor does not attend, the records shall be accepted as accurate.

If the Contractor examines and disagrees the records, and/or does not sign them as agreed, then the Contractor shall give notice to the Engineer of the respects in which the records are asserted to be inaccurate. After receiving this notice, the Engineer shall review the records and either confirm or vary them. If the Contractor does not so give notice to the Engineer within 14 days after being requested to examine the records, they shall be accepted as accurate.

條款 **59**

(續)

香港特別行政區土木工程合同一般條件：

「工程量清單」及計量

(5) (a) 當「本工程」的任何部分需要計量，「工程師」必須通知「承建商」，而「承建商」必須立刻出席或派代表出席協助「工程師」計量，並必須提供「工程師」所要求的一切詳情。如「承建商」不出席或因疏忽或遺漏而沒有派代表出席，則「工程師」的計量會被視為此部分工程的正確計量。

(b) 如要計量須用紀錄及「圖則」方式計量的「永久工程」，「工程師代表」必須為此等工程每月擬備紀錄及「圖則」，而「承建商」必須因應書面要求，在14天內出席以便與「工程師代表」一起審核及同意此等紀錄及「圖則」，並必須在同意後在此等紀錄及「圖則」上簽署。如「承建商」不出席審核及同意任何此等紀錄及「圖則」，這些紀錄及「圖則」會被視為正確的紀錄及「圖則」。

(c) 如「承建商」審核此等紀錄及「圖則」後不同意此等紀錄及「圖則」或經同意後但沒有在此等紀錄及「圖則」上簽署，則除非「承建商」在審核後14天內聲明此等紀錄及「圖則」的不正確之處而向「工程師」呈交一份聲明書，要求「工程師」作出決定，否則這些紀錄及「圖則」亦必須被視為正確的紀錄及「圖則」。

Analysis and application

The operation of this Clause is simple and straightforward. It should be noted that the proper taking of measurement is very important, especially for works which are to be removed or to be hidden from sight like pile lengths, pile obstructions or toe-ins. It is also important to agree records within the time frame stipulated before they can no longer be traced. Taking plenty of Site photographs is a good practice to preserve Site measurement records.

Comparison with the ICE Conditions

The ICE Clause 57 is very similar to the HKSAR Clause 59 but the wording "Unless otherwise provided in the Contract or unless general or detailed description of the work in the Bill of Quantities or other statement clearly shows to the contrary" clearly provides more flexibility and may be less prone to claims arising from the use of the Standard Method of Measurement because any Drawings, Specification or other documents can show a contrary intention and override the procedures in the Standard Method of Measurement. Similar to the HKSAR Clause 59(3), the correction of any error in the description of Bill of Quantities or any omitted item under the ICE Clause 55 is valued by the Engineer as if variations are ordered. The ICE Clause 56(2) makes it an express requirement to consult the Contractor before the Engineer determines an increase or decrease of rate arising from substantial change of quantities in items in the Bill of Quantities.

Comparison with the FIDIC Conditions

In this FIDIC Clause 12.2, there is no reference to any Standard Method of Measurement, which is understandable since the Contract is international. There may be a preamble to the Bill of Quantities, which makes reference to a publication specifying the principles of measurement or describes the principles of measurement for different items of work. Missing items may be discovered during construction and whether they should be measured or not depends on the principles of measurement specified. The FIDIC Clause 12.3 describes the mechanism which triggers re-rating, such as the quantity of an item exceeds 10% of that in the Bill of Quantities, and the change in quantity such that the value of the item exceeds 0.01% of the Accepted Contract Amount, etc.

The FIDIC Clause 12.1 describes how the actual measurement is carried out and is similar to the ICE Conditions and the HKSAR Conditions.

Notes

1. Ian Duncan Wallace, *A Commentary on the FIDIC International Standard Form of Engineering and Building Contract*, p. 119.
2. See the commentary on Clause 56 in Max Abrahamson, *Engineering Law and the ICE Contracts*, 4th Edition, p. 210.
3. Eggleston, *A User's Guide to the ICE Conditions of Contract*, 7th Edition, p. 211.
4. *Henry Boot Construction Ltd. v Alstom Combined Cycles Ltd.* [1999] BLR123.

The HKSAR Conditions:

Variations

(1) The Engineer shall order any variation to any part of the Works that is necessary for the completion of the Works and shall have the power to order any variation that for any other reason shall in his opinion be desirable for or to achieve the satisfactory completion and functioning of the Works. Such variations may include:

    (a) additions, omissions, substitutions, alterations, changes in quality, form, character, kind, position, dimension, level or line;

    (b) changes to any sequence, method or timing of construction specified in the Contract; and

    (c) changes to the Site or entrance to and exit from the Site.

(2) No variation shall be made by the Contractor without an order in writing by the Engineer. No variation shall in any way vitiate or invalidate the Contract but the value of all such variations shall be taken into account in ascertaining the Final Contract Sum.

Corresponding clauses in the ICE Conditions:

Clause 51

(1) The Engineer

    (a) shall order any variation to any part of the Works that is in his opinion necessary for the completion of the Works and

    (b) may order any variation that for any other reason shall in his opinion be desirable for the completion and/or improved functioning of the Works.

    Such variations may include additions omissions substitutions alterations changes in quality form character kind position dimension level or line and changes in any specified sequence method or timing of construction required by the Contract and may be ordered during the Defects Correction Period.

(2) All variations shall be ordered in writing but the provisions of Clause 2(6) in respect of oral instructions shall apply.

(3) No variation ordered in accordance with sub-clause (1) and (2) of this Clause shall in any way vitiate or invalidate the Contract but the value (if any) of all such variation shall be taken into account in ascertaining the amount of the Contract price except to the extent that such variation is necessitated by the Contractor's default.

(4) No order in writing shall be required for increase or decrease in the quantity of any work where such increase or decrease is not the result of an order given under this Clause but is the result of the quantities exceeding or being less than those stated in the Bill of Quantities.

Corresponding clauses in the FIDIC Conditions:

Clause 13.1

Variations may be initiated by the Engineer at any time prior to issuing the Taking-Over Certificate for the Works, either by an instruction or by a request for the Contractor to submit a proposal.

The Contractor shall execute and be bound by each Variation, unless the Contractor promptly give notice to the Engineer stating (with supporting particulars) that the Contractor cannot readily obtain the Goods required for the Variation. Upon receiving this notice, the Engineer shall cancel, confirm or vary the instruction.

Each Variation may include:

(a) changes to the quantities of any item of work included in the Contract (however, such changes do not necessarily constitute a Variation),

(b) changes to the quality and other characteristics of any item of work,

(c) changes to the levels, positions and/or dimensions of any part of the Works,

(d) omission of any work unless it is to be carried out by others,

(e) any additional work, Plant, Materials or services necessary for the Permanent Works, including any associated Tests on Completion, boreholes and other testing and exploratory work, or

(f) changes to the sequence or timing of the execution of the Works.

The Contractor shall not make any alteration and/or modification of the Permanent Works, unless and until the Engineer instructs or approves a Variation.

<table>
<tr><td>

條款
60

</td><td>

香港特別行政區土木工程合同一般條件：

變更命令

</td></tr>
</table>

(1) 為了完成「本工程」的需要，「工程師」必須發出命令對「本工程」的任何部分作出任何變更，及有權命令作出「工程師」因任何其他理由而認為合適的或可使「本工程」圓滿地完成及實現其滿意功能的任何變更。此等變更可包括：

  (a) 增加、刪除、代替、或在質量、形狀、特質、種類、位置、度量、標高或接線方面的改動；

  (b) 「合同」內指定的施工次序、方法或時間的改動；及

  (c) 「工地」或「工地」的出口或入口的改動。

(2) 如沒有「工程師」的書面命令，「承建商」不可作出任何變更。任何變更皆不可在任何情況下使「合同」無效或失效，但在確定「最終合同金額」時必須將所有這些變更的價值一起計算在內。

Commentary

This Clause is relatively short and yet it is one of, if not the most important, clauses in the HKSAR Conditions. The issues arising from this Clause are extremely wide and complex, interconnecting with other subjects such as extension of time, contract risks, pre-contract inspection and sufficiency of Tender, compatibility of Contract documents, method of measurement and valuation, etc. The Clause gives a right to both the Employer and the Contractor. The Employer can, through the Engineer, orders variations necessary for the completion of the Works or desirable for or to achieve satisfactory completion and functioning of the Works. Without the Clause, the Contractor is not obliged to do more or different work than agreed in the original Contract, nor is the Contractor obliged to accept an omission of the work. The Engineer also derives his power to order variation relying on this Clause, since he has no ostensible authority to contract any additional or extra work on behalf of the Employer. In practice, in most HKSAR contracts, the Employer limits the power of the Engineer to order variation up to $300,000 without the consent of the Employer. However, this limit appears illusionary since for many major contracts, the necessary extension of time in a variation attracts prolongation cost which could be in the order of hundreds of thousands dollars a day, and therefore the Engineer must consider the time consequence when ordering any variation. As far as the Contractor is concerned, the Clause entitles the Contractor to receive a variation order from the Engineer if the variation is necessary for the completion of the Works. For example, the Contractor encounters construction difficulties which the Employer or the Engineer is responsible such as faulty design or incorrect Specification.

A lot of arguments arise from whether any specific works are included in the original Contract or in fact variations. If the Engineer wrongly orders any "variation," which is actually included in the original Contract, it will usually not bind the Employer.[1] For example, in piling work, there may be an argument regarding whether any increase in the pile length due to unexpected rockhead level is a variation or not. This is because if the Contractor successfully gets a variation order, he will be entitled to a varied rate and more importantly, extension of time with subsequent prolongation cost. It is submitted that, in answering the question, the Contract must be read as a whole. Wording like "to be determined on Site by the Engineer" or "to be proposed by the Contractor and agreed by the Engineer" in the Specification or Drawings are relevant and therefore those responsible for preparing the design Drawings and Specification should be well aware of the consequences. Similarly, typical Drawings for slope profile must be carefully considered with respect to unexpected ground conditions, for example, difference in rockhead levels could subsequently necessitate a change in slope profile and therefore flexibility should be allowed in defining the typical details to avoid claims for variation.

Further arguments may also arise as to whether any instruction from the Engineer falls within the definition of variation or in fact outside the terms of the Contract, especially with respect to any omission of the Works. If the Contractor successfully argues the case, the Employer would be obliged to sign a supplementary agreement with the Contractor who may demand terms different from those set out in the original Contract if the Employer wishes the additional work to be carried out or any original work to be omitted from the Contract.

Analysis and application

Clause 60(1) "The Engineer shall order any variation to any part of the Works that is necessary for the completion of the Works ..."

This first part of this sub-clause (1) makes it an obligation for the Engineer to order variations to the Works if it is necessary for completion of the Works. For example, if the Contract specifies the use of a material which has become unavailable in the market, the Engineer must change to another substitute material which is available by issuing a variation order to complete the specific part of the Works or if there are design deficiencies, the Engineer must rectify the design and issue new Drawings to the Contractor, which is also a variation to the Contract. Please note that the test on the validity of a variation is the wording "necessary for the completion of the Works." The obligation and power rest upon the Engineer and not the Employer and if it is necessary during construction for the Engineer to order additional safety means to complete a specific part of the Works at an additional cost, the Employer cannot object. Sometimes, the Employer may like to add major additional Works to the Contract, for example, building an additional house when the market condition is good. This would change the scope of the Contract and is not actually necessary for the completion of the Works and would therefore fall outside the provision of this Clause 60. To do this, the Employer needs to sign a Supplementary Agreement with the Contractor with rates and other terms to be agreed upon between the Parties.

Clause 60(1) "... and shall have the power to order any variation that for any other reason shall in his opinion be desirable for or to achieve satisfactory completion and functioning of the Works ..."

This second part of sub-clause (1) gives the Engineer the power, not an obligation, to order any variation he considers desirable for or to achieve satisfactory completion and functioning of the Works. The scope of variation is wider in the sense that the Engineer's opinion is taken into consideration in deciding whether any instruction falls within the definition of variation. Also, apart from completing the Works, the scope of variation also covers the satisfactory functioning of the Works. For example, the variation can include any improvement by adding additional components to a water treatment plant if the Engineer considers it desirable to do so. Another example is that changing the surface protection to a steel structure from painting to metal coating falls within the Engineer's power since it is desirable for the satisfactory functioning of the Works. On the other hand, changing to a different colour of a building may, strictly speaking, not fall within the Engineer's power. For this second part of sub-clause (1), it is very often that the Engineer would consult the Employer as regards to his requirements or it may arise from a request of the Employer. In such case, the variation should be valid if the Engineer considers it to be desirable for the satisfactory completion and functioning of the Works.

Clause 60(1)(a) "additions, omissions, substitutions, ... changes in quality, form, character, kind, ..."

It is suggested that changes in quality, form, character, or kind, etc. of the whole of the Works can never be desirable for the satisfactory completion of the Works because the Works must mean the original scope of the Works when the Contract was formed.[2] It is also suggested that large omissions are unlikely to be necessary or desirable for the satisfactory completion and functioning of the Works.[3] Omission would also entails in loss of profit which always attract arguments as to whether it could be recoverable and apparently the HKSAR Conditions might not have allowed for it if the omission falls within the definition of variation.[4]

Clause 60(1)(b) "changes to any sequence, method or timing of construction specified in the Contract; ..."

The changes to the specified timing of construction are not appropriate to cover an order to complete the whole Works before the Contract completion date.[5] The variation order may cover changes to the sequence and method of construction specified in the Contract. An example is that the traffic stagings are shown on the Drawings and during the construction period, these stagings are not followed. This may give rise to a variation order if the alternative stagings are necessary to suit prevailing Site conditions. Likewise, if the Drawing shows the method and sequence of construction of a segmental bridge deck construction, any Site condition which requires a change of construction method and/or sequence may give rise to a variation.

Clause 60(1)(c) "changes to the Site or entrance to and exit from the Site."

This may apply to changes in the boundary of the Site or the physical conditions of the Site, which affect the Works.

Clause 60(2) "… No variation shall in any way vitiate or invalidate the Contract but the value of all such variations shall be taken into account in ascertaining the Final Contract Sum."

The Contractor must comply with all valid variation orders, for example, numerous orders to rectify the design deficiencies in the Works but then he is entitled to all the cost of the varied works including disruption cost and prolongation cost associated with any extension of time.

Comparison with the ICE Conditions

The sub-clause (1), (2) and (3) of this ICE Clause is very similar to the HKSAR Clause 60. The wording "improved functioning of the Works" gives the Engineer a wider power than just "satisfactory functioning of the Works" allowed for in the HKSAR Clause. The ICE Clause also allows for variation to be ordered during the Defects Correction Period. In the HKSAR contracts, a Special Condition allows for this. The ICE Clause allows oral instructions to order variation provided that it is confirmed in writing as soon as possible under the circumstance. Sub-clause (4) states explicitly that increase and decrease in quantities as a result of inaccurate estimate or actual re-measurement on Site requires no variation order from the Engineer.

Comparison with the FIDIC Conditions

There are essential differences in this FIDIC Clause as against the HKSAR Clause and the ICE Clause. The Contractor is entitled to give notice to the Engineer if he cannot readily obtain the Goods (which are the Contractor's Equipment, Materials, Plant and Temporary Works) required for the variation. If notice is given, the notice and supporting particulars should be carefully studied. If the variation is confirmed and not varied, the confirmation should be in writing and address the issues raised in the Contractor's notice. If the variation itself is varied, it may be a new variation.[6] Another substantial difference is that sub-paragraph (f) allows changes to the sequence or timing of the execution of the Works. It is submitted that the absence of the word "specified" gives a much wider scope for the Engineer to change the sequence and timing of the Works, even if the sequence and timing are chosen by the Contractor, for example, by ordering the Contractor to mitigate delay which otherwise the Engineer may need to give extension of time.

Notes

1. Ian Duncan Wallace, *A Commentary on the FIDIC International Standard Form of Engineering and Building Contract*, p. 95.
2. Max Abrahamson, *Engineering Law and the ICE Contracts*, 4th Edition, p. 173.
3. Max Abrahamson, *Engineering Law and the ICE Contracts*, 4th Edition, p. 173 (together with several examples of changes falling outside the ICE variation clause which is almost identical to the HKSAR variation clause).
4. For a general discussion on compensation on loss of profit arising from variation, see Roger Knowles, *150 Contractual Problems and Their Solutions*, Sections 8.3 and 8.4.
5. Max Abrahamson, *Engineering Law and the ICE Contracts*, 4th Edition, p. 176.
6. Peter L. Booen (Principle drafter), *The FIDIC Contracts Guide*.

The HKSAR Conditions:

Valuing Variations

(1) The Engineer shall determine the sum which in his opinion shall be added to or deducted from the Contract Sum as a result of an order given by the Engineer under Clause 60 in accordance with the following principles:

    (a) Any item of work omitted shall be valued at the rate set out in the Contract for such work.

    (b) Any work carried out which is the same as or similar in character to and executed under the same or similar conditions and circumstances to any item of work priced in the Contract shall be valued at the rate set out in the Contract for such item of work.

    (c) Any work carried out which is not the same as or similar in character to or is not executed under the same or similar conditions or circumstances to any item of work priced in the Contract shall be valued at a rate based on the rates in the Contract so far as may be reasonable, failing which, at a rate agreed between the Engineer and the Contractor.

Corresponding clauses in the ICE Conditions:

Clause 52

(1) If requested by the Engineer the Contractor shall submit his quotation for any proposed variation and his estimate of any consequential delay. Wherever possible the value and delay consequences (if any) of each variation shall be agreed before the order is issued or before work starts.

(2) Where a request is not made or agreement is not reached under sub-clause (1) the valuation of variations ordered by the Engineer in accordance with Clause 51 shall be ascertained as follows.

    (a) As soon as possible after receipt of the variation the Contractor shall submit to the Engineer

        (i) his quotation for any extra or substituted works necessitated by the variation having due regard to any rates or prices included in the Contract and

        (ii) his estimate of any delay occasioned thereby and

        (iii) his estimate of the cost of any such delay.

    (b) Within 14 days of receiving the said submissions the Engineer shall

        (i) accept those submissions or

        (ii) negotiate with the Contractor thereon.

    (c) Upon reaching agreement with the Contractor the Contract Price shall be amended accordingly.

(3) Failing agreement between the Engineer and the Contractor under either sub-clause (1) or (2) the value of variations ordered by the Engineer in accordance with Clause 51 shall be ascertained by the Engineer in accordance with the following principles and be notified to the Contractor.

    (a) Where work is of similar character and carried out under similar conditions to work priced in the Bill of Quantities it shall be valued at such rates and prices contained therein as may be applicable.

    (b) Where work is not of a similar character or is not carried out under similar conditions or is ordered during the Defects Correction Period the rates and prices in the Bill of Quantities shall be used as the basis for valuation so far as may be reasonable failing which a fair valuation shall be made.

Corresponding clauses in the FIDIC Conditions:

Clause 13.3

If the Engineer requests a proposal, prior to instructing a Variation, the Contractor shall respond in writing as soon as practicable, either by giving reasons why he cannot comply (if this is the case) or by submitting:

(a) a description of the proposed work to be performed and a programme for its execution,

(b) the Contractor's proposal for any necessary modifications to the programme according to Sub-Clause 8.3 [*Programme*] and to the Time for Completion, and

(c) the Contractor's proposal for evaluation of the Variation.

The Engineer shall, as soon as practicable after receiving such proposal (under Sub-Clause 13.2 [*Value Engineering*] or otherwise), responds with approval, disapproval or comments. The Contractor shall not delay any work whilst awaiting a response.

Each instruction to execute a Variation, with any requirements for the recording of Costs, shall be issued by the Engineer to the Contractor, who shall acknowledge receipt.

Each Variation shall be evaluated in accordance with Clause 12 [*Measurement and Evaluation*], unless the Engineer instructs or approves otherwise in accordance with this Clause.

Clause 12.3

Except as otherwise stated in the Contract, the Engineer shall proceed in accordance with Sub-Clause 3.5 [*Determinations*] to agree or determine the Contract Price by evaluating each item of work, applying the measurement agreed or determined in accordance with the above Sub-Clauses 12.1 and 12.2 and the appropriate rate or price for the item.

For each item of work, the appropriate rate or price for the item shall be the rate or price specified for such item in the Contract or, if there is no such item, specified for similar work.

Clause 3.5

Whenever these Conditions provide that the Engineer shall proceed in accordance with this Sub-Clause 3.5 to agree or determine any matter, the Engineer shall consult with each party in an endeavour to reach agreement. If agreement is not achieved, the Engineer shall make a fair determination in accordance with the Contract taking regard of all circumstances.

條款 61

香港特別行政區土木工程合同一般條件：

對變更釐定價值

(1) 「工程師」根據第60條發出命令之後，他必須按下列原則訂立他認為必須在「合同款項」內予以增加或扣除的款項：

    (a) 任何被刪除的工程項目必須用「合同」內所列明的這項工程價率釐定其價值。

    (b) 任何與「合同」內所標價的工程項目特質相同或類似，並在相同或類似的條件及情況下所實施的工程，必須用「合同」內所列出的價率釐定其價值。

    (c) 任何與「合同」內所標價的工程項目特質不同或不類似，或並非在相同或類似的條件或情況下所實施的工程，必須在合理的情況下按照「合同」的價率釐定其價值。如沒有適當價率，則用「工程師」與「承建商」之間所同意的價率釐定其價值。

Commentary

This Clause sets out the principle of valuation of works not merely arising from variation orders under Clause 60 but also other reasons such as clarification of ambiguities and discrepancies of documents under Clause 5, investigating defects under Clause 58, as well as errors and omissions in the Bill of Quantities under Clause 59. Its application is actually wider because other works such as a direction for variation of unsatisfactory material and work under Clause 46(3) and execution of work of repair under Clause 56(3) are also regarded as variations. The principles set out in sub-clause (a), (b) and (c) applicable to omission of work, additional work of a similar character executed under similar conditions, and additional work not of similar character or executed under similar conditions respectively appear straightforward, but in practice it gives rise to numerous claims and disputes, especially for items of work which have been speculatively over-priced or under-priced due to incorrect or inaccurate quantities in the Bill of Quantities. The proviso to sub-clause (1) regarding the determination of a new rate when the nature and extent of variation render the original Contract rates unreasonable and inapplicable also gives rise to a lot of disputes resulting in arbitration or litigation.

Analysis and application

Clause 61(1) "The Engineer shall determine the sum … as a result of an order given by the Engineer under Clause 60 …"

    Under the HKSAR Conditions, virtually every change of details in Drawings or Specification is regarded as a variation. The exception to it is when the Drawing details provide for flexibility to allow for variable Site conditions using the wording such as "to be determined on Site by the Engineer" and "levels and alignments are tentative only subject to Engineer's determination," etc. As explained in the commentary to Clause 60, allowing for such flexibility is essential and this approach not only avoids variations with resulting extension of time, varied rates and prolongation costs, but also a lot of administration work in numerous valuations and fixing rates by the Engineer. If variations are indeed required, the Engineer must value the variations as soon as possible and determine the rates and prices resulting in a sum to be added to or deducted from the Contract Sum.

Clause 61(1)(a) "Any item of work omitted shall be valued at the rate set out in the Contract for such work."

This sub-clause applies to omission of work through variations and the value of work is simply based on the Contract rate of the particular item to be omitted. It does not apply to valuation of the missing items under Clause 59(3) since there may not be an existing rate under the Contract. Sometimes, there may be work or services which are required under the Contract stipulated in Specification or in drawings and which may be covered under the general responsibility or obligation. In such case, the clause will not be applicable and the reduction of Cost due to any omission of such work or service may have to be agreed between the Engineer and the Contractor.

Clause 61(1)(b) "Any work carried out which is the same as or similar in character to and executed under the same or similar conditions and circumstances to any item of work priced in the Contract shall be valued at the rate set out in the Contract for such item of work."

This sub-clause is straightforward in the sense that additional work of similar character executed under similar condition to the work of which a rate is priced by the Contractor shall be valued at that rate. Even if there is a mistake in pricing or any speculative pricing resulting in abnormally high or low rate, the rate shall apply.[1] This principle applies equally to substantial increase or decrease in quantity due to incorrect quantities in the Bill of Quantities unless there is a change in the method of construction or in the economics of the construction.

Clause 61(1)(c) "Any work carried out which is not the same as or similar in character ... shall be valued at a rate based on the rates in the Contract so far as may be reasonable, ..."

For additional work not of similar character executed under similar conditions to the work with a priced rate, the proper way to value the additional work is to produce a break-down into the cost components of the work based on the existing rates and then build-up the new rate for the dissimilar work or work executed under a different condition from the aforesaid cost components, taking account of the particular character and working conditions of the additional work. For example, deeper piles involve more difficulty in excavation, keeping water out and concreting. The wording "so far as may be reasonable" needs some careful interpretation with reference to Court cases. The price or rate mistakenly entered high or low by the Contractor may not be considered by the valuer as being unreasonable based on Henry Boot's case but factors like the varied work (type B door) has to be obtained from a named supplier in the Contract at a price much higher than a build up of rate from the Contract rate (calculated based on the quotation from the aforesaid named supplier for type A door) must be taken into account.[2]

 34   In an arbitration on the valuation of the Cost of an addition ramp, is this model useful?

It is recently decided that a section of parapet 200 metres long has to be constructed up to 1.5 metres high instead of 1.2 metres high. Describe how the Engineer values such change.

The HKSAR Conditions:

Valuing Variations

Provided that if the nature or extent of any variation ordered in accordance with Clause 60 relative to the nature or extent of the Works or any part thereof shall be such that in the opinion of the Engineer any rate contained in the Contract for any item of work is by reason of such variation rendered unreasonable or inapplicable then a new rate shall be agreed between the Engineer and the Contractor for that item, using the Contract rates as the basis for determination.

(2) In the event of the Engineer and the Contractor failing to reach agreement on any rate under the provisions of sub-clause (1) of this Clause, the Engineer shall fix such rate as shall in his opinion be reasonable and notify the Contractor accordingly.

Corresponding clauses in the ICE Conditions:

Clause 52 (*continued*)

(4) If in the opinion of the Engineer or the Contractor any rate or price contained in the Contract for any item of work (not being the subject of any variation) is by reason of any variation rendered unreasonable or inapplicable either the Engineer shall give to the Contractor or the Contractor shall give to the Engineer notice before the varied work is commenced or as soon thereafter as is reasonable in all the circumstances that such rate or price should be increased or decreased and the Engineer shall fix such rate or price as in the circumstances he shall think reasonable and proper and so notify the Contractor.

Corresponding clauses in the FIDIC Conditions:

Clause 12.3 (*continued*)

However, a new rate or price shall be appropriate for an item of work if:

(a) (i) the measured quantity of the item is changed by more than 10% from the quantity of this item in the Bill of Quantities or other Schedule,

(ii) this change in quantity multiplied by such specified rate for this item exceeds 0.01% of the Accepted Contract Amount,

(iii) this change in quantity directly changes the cost per unit quantity of this item by more than 1%, and

(iv) this item is not specified in the Contract as a "fixed rate item";

or

(b) (i) the work is instructed under Clause 13 [*Variations and Adjustments*],

(ii) no rate or price is specified in the Contract for this item, and

(iii) no specified rate or price is appropriate because the item of work is not of similar character, or is not executed under similar conditions, as any item in the Contract.

Each new rate or price shall be derived from any relevant rates or prices in the Contract, with reasonable adjustments to take account of the matters described in sub-paragraph (a) and/or (b), as applicable. If no rates or prices are relevant for the derivation of a new rate or price, it shall be derived from the reasonable cost of executing the work, together with reasonable profit, taking account of any other relevant matters.

Until such time as an appropriate rate or price is agreed or determined, the Engineer shall determine a provisional rate or price for the purposes of Interim Payment Certificates.

Notes

1. *Henry Boot Construction Ltd. v Alstoms Combined Cycles Ltd.* [1999] BLR123.
2. *Hong Kong Housing Authority v Leighton Contractors (Asia) Ltd.* (2004).
3. *Government Practice Note on GCC61 and 63(b).* See also the commentary to Clause 52(2) in Max Abrahamson, *Engineering Law and the ICE Contracts*, 4th Edition, p. 188, n. 12.
4. *Weldon Plant v Commissioner for New Towns* [2000] TCCBLR496; *Tinghamgrange v Dew* (1995) 47 con LR105. For a general discussion about delay costs arising from variation, see Roger Knowles, *150 Contractual Problems and Their Solutions*, Section 8.1.

對變更釐定價值

（續）

但是，如根據第60條所發出的任何工程變更命令的性質或程度相對於「本工程」或其任何部分的性質或程度使「工程師」認為「合同」內所載任何工程項目的價率由於此變更變得不合理或不適用，則「工程師」與「承建商」必須用「合同」內的價率作基礎訂立一個雙方同意的新價率。

(2) 如「工程師」與「承建商」未能按照本條第(1)款的條文為任何價率達成協議，「工程師」必須定出他認為合理的價率並就此通知「承建商」。

Analysis and application

Clause 61(1) "Provided that if the nature or extent of any variation … relative to the nature or extent of the Works … in the opinion of the Engineer any rate … is by reason of such valuation rendered unreasonable or inapplicable then a new rate shall be agreed between the Engineer and the Contractor … using the Contract rate as the basis for determination."

For valuation under sub-clause (1)(a), (b) and (c), the rate to be valued should refer to the varied works. For this sub-clause, the wording "any rate" should then refer to the unvaried works.[3] If the Engineer changes the conditions and circumstances of the unvaried works by ordering variation, the Contractor is entitled to an appropriate adjustment for any knock-on effect of the unvaried works, for example, buying additional insurance, maintaining traffic, Site expenses, etc. Some commentators also say that prolongation costs resulting from delay in overall progress of the Works could also be valued under Clause 61 instead of under Clause 63(b) and the rate or rates should be adjusted by adding such costs. It is submitted that such suggestion may give rise to a lot of ambiguities. Firstly, the rates would be artificially pumped up to many times of the original rates. Secondly, different results entail from valuation under Clause 61 and valuation under Clause 63(b) because the former includes profit.[4] Thirdly, the valuation of variation order is the duty of the Engineer. If prolongation costs are valued under Clause 61 and the Engineer is unaware that the proposed variation has time implication, the Contractor can introduce claims through the back door without complying with the requirements of giving notice and particulars under Clause 64. The barring of the Contractor's out of notice claims may not be possible and the Employer's right would be seriously prejudiced.

Clause 61(2) "… failing to reach agreement on any rate … the Engineer shall fix such rate …"

This allows the Engineer to fix a rate and pay the Contractor accordingly until a new rate is determined by the Engineer after the Contractor raises a claim under Clause 64 within 28 days.

Comparison with the ICE Conditions

The ICE sub-clauses (1) and (2) detail the procedures which the Engineer should follow, with a view to agreeing on the valuation of variation ordered or to be ordered by the Engineer. The Contractor is given an opportunity to submit his quotation, including any delay consequences. The emphasis is on reaching a consensus on the value of variation instead of leaving the valuation aside for a long time, resulting in hundreds of variations yet to be valued towards the end of the Contract period. If consensus is not reached, a set of principles very similar to the HKSAR Conditions is to be used by the Engineer to value the variation unilaterally.

Comparison with the FIDIC Conditions

The FIDIC Clause 13.3 allows the Contractor to submit a proposal before instructing a variation, describing the proposed work and the programme for its execution as well as the proposal for the evaluation of the variation. The time for approval of the Contractor's proposal shall not delay the work. Clause 12.3 sets out the principles for evaluation. Basically, the appropriate rate for the additional work shall be the rate specified for the item corresponding to the additional work or if there is no corresponding item, the relevant rates in similar items. For works with no relevant rates in the Contract, the rate shall be derived from reasonable Cost with reasonable profit. Clause 3.5 requires the Engineer to consult each party in an endeavour to reach agreement, leaving a wider scope for consultation and adjustment of rate than the HKSAR Clause.

(Please see opposite page for Notes.)

Clause 62

The HKSAR Conditions:

Daywork

(1) The Engineer may, if in his opinion it is necessary or desirable, order in writing that any work to be carried out as a result of a variation ordered under Clause 60 shall be executed on a daywork basis.

(2) The Contractor shall then be paid for such work under the conditions and at the rates set out in the Contract or if no such conditions and rates have been included, at such rates as the Engineer shall determine as being reasonable.

(3) The Contractor shall furnish to the Engineer such receipts or other vouchers as may be necessary to prove the sums paid and before ordering materials shall, if so required by the Engineer, submit to the Engineer quotations for the same for his approval.

(4) In respect of all work executed on a daywork basis the Contractor shall during the continuance of such work deliver each working day to the Engineer's Representative a list, in duplicate, of the names and occupations of and time worked by all workers employed on such work on the previous working day and a statement, also in duplicate, showing the descriptions and quantity of all materials and Constructional Plant used thereon or therefor. One copy of such lists and statements shall be agreed as correct or be rejected with stated reasons, be signed by the Engineer's Representative and returned to the Contractor within 2 days exclusive of General Holidays.

(5) At the end of each month the Contractor shall deliver to the Engineer's Representative a priced statement of the labour, materials and Constructional Plant used on a daywork basis.

Provided that if the Engineer shall consider that for any reason the sending of such statement by the Contractor in accordance with the foregoing provision was impracticable the Engineer shall nevertheless be entitled to authorize payment for such work either as daywork, on being satisfied as to the time employed and the Constructional Plant and materials used thereon, or at such value as shall in the Engineer's opinion be reasonable.

(6) The Contractor shall inform the Engineer's Representative in advance whenever the Contractor proposes to carry out daywork ordered by the Engineer and shall afford every facility for the Engineer's Representative to check all time and materials for which the Contractor proposes to charge therefor.

Corresponding clauses in the ICE Conditions:

Clause 52

(5) The Engineer may if in his opinion it is necessary or desirable order in writing that any additional or substituted work shall be carried out on a daywork basis in accordance with provisions of Clause 56(4).

Clause 56

(4) Where any work is carried out on a daywork basis the Contractor shall be paid for such work under the conditions and at the rates and prices set out in the daywork schedule included in the Contract or failing the inclusion of a daywork schedule he shall be paid at the rates and prices and under the conditions contained in the "Schedules of Daywork carried out incidental to Contract Work" issued by The Civil Engineering Contractors Association (formerly issued by the Federation of Civil Engineering Contractors) current at the date of the carrying out of the daywork.

The Contractor shall furnish to the Engineer such records receipts and other documentation as may be necessary to prove amounts paid and/or costs incurred. Such returns shall be in the form and delivered at the times the Engineer shall direct and shall be agreed within a reasonable time.

Before ordering materials the Contractor shall if so required submit to the Engineer quotations for the same for his approval.

Corresponding clauses(s) in the FIDIC Conditions:

Clause 13.6

For work of a minor or incidental nature, the Engineer may instruct that a Variation shall be executed on a daywork basis. The work shall then be valued in accordance with the Daywork Schedule included in the Contract, and the following procedure shall apply. If a Daywork Schedule is not included in the Contract, this Sub-Clause shall not apply.

Before ordering Goods for the work, the Contractor shall submit quotations to the Engineer. When applying for payment, the Contractor shall submit invoices, voucher and accounts or receipts for any Goods.

Except for any items for which the Daywork Schedule specifies that payment is not due, the Contractor shall deliver each day to the Engineer accurate statements in duplicate which shall include the following details of the resources used in executing the previous day's work:

(1) the names, occupations and time of Contractor's Personnel,

(2) the identification, type and time of Contractor' Equipment and Temporary Works, and

(3) the quantities and types of Plant and Materials used.

One copy of each statement will, if correct, or when agreed, be signed by the Engineer and returned to the Contractor. The Contractor shall then submit priced statements of these resources to the Engineer, prior to their inclusion in the next Statement under Sub-Clause 14.3 [*Application for Interim Payment Certificates*].

條款 62

香港特別行政區土木工程合同一般條件：

日工

(1) 如「工程師」認為有需要或應當的話，他可書面命令任何根據第60條所發出的變更命令所施行的工程以日工方式實施。

(2) 「承建商」則須依「合同」內所規定的條件及所列出的價率收取此項工程的付款，如「合同」並無包括此等條件及價率，「承建商」則須以「工程師」釐定為合理的價率收取付款。

(3) 「承建商」須提交給「工程師」所需的收據或其他單據證明已獲支付的款項，如「工程師」要求，「承建商」亦必須在訂購物料之前向「工程師」提交物料的報價單給他批准。

(4) 對所有以日工方式實施的工程，「承建商」必須在此工程持續期間的每個工作日向「工程師代表」提交一式兩份的清單，列出上個工作日受僱於此項工程的所有工人的姓名、職業及工作時間，及一式兩份的報表，列出用於此項工程或為此項工程而使用的一切物料及「施工設備」的詳情及數量。以上的清單及報表的一份印本必須由「工程師代表」同意為正確或以他所陳述的理由而被他拒絕接納，並由他簽署及在兩天內(公眾假期除外)交還給「承建商」。

(5) 「承建商」須在每月月尾提交給「工程師代表」一份以日工方式施行的工程所用的工人、物料及「施工設備」的已標價的報表。

但是，如「工程師」以任何理由認為「承建商」按以上的條文提交報表並不實際可行，「工程師」有權允許按日工標準為有關工程付款，或若「工程師」同意僱用的時間及所用的「施工設備」及物料，「工程師」可按他認為合理的價格付款。

(6) 每當「承建商」提議施行「工程師」所命令施行的日工時，他必須事前通知「工程師代表」，並須盡力提供一切設施，以給「工程師代表」查閱「承建商」所提議而因此收費的所有時間及物料。

Commentary

In Hong Kong, daywork is usually employed for work of a minor nature, such as fencing an area of the Site where no allowance has been made in the Contract for such work. Strictly speaking, ordering daywork is a variation but since there are Bill items allowing payment for such work, daywork can be regarded as works of provisional nature subject to the Engineer's discretion to order or not. For this Clause 62, except for sub-clauses (1) and (2), the other sub-clauses are on procedural matters and are straightforward.

Analysis and application

Clause 62(1) "The Engineer may, if in his opinion it is necessary or desirable, order in writing that any work to be carried out as a result of a variation ordered under Clause 60 shall be executed on a daywork basis."

Though the wording appears to have given the Engineer wide discretion to order daywork which should normally be carried out under a variation order, it should only be ordered where a valuation at other rates is not practical but not as a means to deprive the Contractor of the benefit of high rates or the burden of low rates on which the valuation of the variation would otherwise be based.[1]

Clause 62(2) "The Contractor shall then be paid for such work under the conditions and at the rate set out in the Contract or if no such conditions and rates have been included, at such rate as the Engineer shall determine as being reasonable."

Since the rates in the daywork schedule are usually priced very low because the tenderers foresee that daywork would not be ordered a lot, the Contractor would normally resist using daywork rates as a basis for valuing variations. For labour cost at least, the Contractor prefers using the Average Daily Wages of Workers published by the Census and Statistical Department, which usually gives a higher rate. It is submitted that daywork should only be ordered for simple and straightforward works, and if the works are complicated and require a lot of supervisory and engineering input, daywork rates may not be appropriate and a proper variation order valued on the basis of Clause 61 would be more appropriate.

Comparison with the ICE Conditions

These ICE sub-clauses 52(5) and 56(4) are effectively the same as the HKSAR Clause 62 except that the "Schedule of Daywork carried out incidental to Contract Works" are specified to be used where no daywork schedule is included in the Contract. The procedural description is much more general and shorter.

Comparison with the FIDIC Conditions

This FIDIC Clause 13.6 is similar to the ICE Clause and the HKSAR Clause except that it explicitly mentions that daywork is used for work of a minor or incidental nature only and this actually reflects the reality.

Notes

1. Max Abrahamson, *Engineering Law and the ICE Contracts*, 4th Edition, p. 190.

 This photo shows the control centre for a tunnel. There are several electrical components missing from the Drawings and the Bill of Quantities. Can the Engineer uses daywork to install these components?

Q 37 A dangerous boulder on the natural slope has to be removed. Should the Engineer order daywork to remove it? Who is responsible if during the course of removal, the boulder falls down and damages the falework of the flyover construction?

Clause 63

Disturbance to the Progress of the Works

If upon written application by the Contractor to the Engineer, the Engineer is of the opinion that the Contractor has been or is likely to be involved in expenditure for which the Contractor would not be reimbursed by a payment made under any other provision in the Contract by reason of the progress of the Works or any part thereof having been materially affected by:

(a) the Contractor not having received in due time necessary instructions, orders, directions, decisions, Drawings, Specification, details or levels from the Engineer for which the Contractor specifically applied in writing on a date which, having regard to the time for completion of the Works prescribed by Clause 49 or to any extension of time then granted by the Engineer, was neither unreasonably distant from nor unreasonably close to the date on which it was necessary for the Contractor to receive the same, or

(b) any variation ordered in accordance with Clause 60, or

(c) the opening up for inspection in accordance with Clause 45 of any work covered up or the testing of materials or workmanship not required by the Contract but directed by the Engineer or the Engineer's Representative in accordance with Clause 42(1) unless the inspection or test showed that the work, materials or workmanship were not in accordance with the Contract, or

(d) delay caused by any person or any company, not being a utility undertaking, engaged by the Employer in supplying materials or in executing work directly connected with but not forming part of the Works, or

(e) late delivery of material, Plant or equipment by the Employer,

then the Engineer shall ascertain the Cost incurred and shall certify in accordance with Clause 79.

Corresponding clauses in the ICE Conditions:

Clause 7

(4) If by reason of any failure or inability of the Engineer to issue at a time reasonable in all the circumstances Drawings Specifications or instructions requested by the Contractor and considered necessary by the Engineer in accordance with sub-clause (1) of this Clause the Contractor suffers delay or incurs additional cost then the Engineer shall take such delay into account in determining any extension of time to which the Contractor is entitled under Clause 44 and the Contractor shall subject to Clause 53 be paid in accordance with Clause 60 the amount of such cost as may be reasonable.

Clause 27

(2) (b) Any condition or limitation in any licence obtained after award of the Contract shall be deemed to be an instruction under Clause 13.

(3) The Contractor shall be responsible for giving to any relevant authority any required notice (or advance notice where prescribed) of his proposal to commence any works. A copy of each such notice shall be given to the Employer.

(4) If any instruction pursuant to sub-clause (2)(b) of this Clause results in delay to the construction and completion of the Works because the Contractor needs to comply with sub-clause (3) of this Clause the Engineer shall in addition to valuing the variation under Clause 52 take such delay into account in determining any extension of time to which the Contractor is entitled under Clause 44 and the Contractor shall subject to Clause 53 be paid in accordance with Clause 60 such additional cost as the Engineer shall consider to have been reasonably attributable to such delay.

Clause 38

(2) The Contractor shall uncover any part or parts of the Works or make openings in or through the same as the Engineer may from time to time direct and shall reinstate and make good such part or parts to the satisfaction of the Engineer. If any such part or parts have been covered up or put out of view after compliance with the requirements of sub-clause (1) of this Clause and are found to have been carried out in accordance with the Contract the cost of uncovering making openings in or through reinstating and making good the same shall be borne by the Employer but in any other case all such cost shall be borne by the Contractor.

Corresponding clauses in the FIDIC Conditions:

Clause 1.9

If the Contractor suffers delay and/or incurs Cost as result of a failure of the Engineer to issue the notified drawing or instruction within a time which is reasonable and is specified in the notice with supporting details, the Contractor shall give a further notice to the Engineer and shall be entitled subject to Sub-Clause 20.1 [*Contractor's Claims*] to:

(a) an extension of time for any such delay, if completion is or will be delayed, under Sub-Clause 8.4 [*Extension of Time for Completion*], and

(b) payment of any such Cost plus reasonable profit, which shall be included in the Contract Price.

Clause 8.5

If the following conditions apply, namely:

(a) the Contractor has diligently followed the procedures laid down by the relevant legally constituted public authorities in the Country,

(b) these authorities delay or disrupt the Contractor's work, and the delay or disruption was unforeseeable,

then this delay or disruption will be considered as a cause of delay under sub-paragraph (b) of Sub-Clause 8.4 [*Extension of Time for Completion*].

Clause 11.8

The Contractor shall, if required by the Engineer, search for the cause of any defect, under the direction of the Engineer. Unless the defect is to be remedied at the cost of the Contractor under Sub-Clause 11.2 [*Cost of Remedying Defects*], the cost of the search plus reasonable profit shall be agreed or determined by the Engineer in accordance with Sub-Clause 3.5 [*Determinations*] and shall be included in the Contract Price.

條款 63

香港特別行政區土木工程合同一般條件：

「本工程」進展受阻礙

如「承建商」向「工程師」提出書面申請而「工程師」認為「本工程」或其任何部分的進展受到下列任何事項的實質影響導致「承建商」已經或很大機會會牽涉一些支出，而「承建商」不會從「合同」內的任何其他條文獲得付還，則「工程師」必須確定已引致的成本並根據第79條予以核實：

(a) 「承建商」未能在適當時間內從「工程師」獲得所需的指令、命令、指示、決定、「圖則」、規格、詳情或標高，而「承建商」發出書面申請上述文件的日期，在顧及到按照第49條訂明的「本工程」竣工的時限或任何由「工程師」給予的延期後，距離「承建商」有需要收到上述文件的日期並非不合理地太遠或太近；或

(b) 根據第60條發出的任何變更命令；或

(c) 「承建商」按照第45條規定打開任何已被遮蓋的工程以供檢查或按照第42條(1)款在「工程師」或「工程師代表」指示下進行「合同」內並沒有規定的物料或技術水平測試，除非此項檢查或測試顯示這些工程、物料或技術水平並不符合「合同」的規定；或

(d) 由任何人或公司（公共事業機構除外）造成的延誤；這些人或公司是由「僱主」聘用以供應物料，或實施與「本工程」有直接關連但不構成「本工程」任何部分的工程；或

(e) 「僱主」逾期交予物料、機械或器材。

Commentary

This Clause 63 allows for reimbursement of Cost incurred by reason of a disturbance to the progress of the Works. For this Clause to operate, the disruptive event needs not affect the overall critical activities such that the Contractor is entitled to an extension of time. The disruptive effect can be localized idling of labour, Plant, deterioration of materials or extra management input without affecting any sectional or overall completion of the Works. The most common type of claims under this Clause arises from late instruction (sub-clause (a)) and/or variation (sub-clause (b)) and it has been quite common for the Contractor to lump all such effects together and submit a global claim,[1] which has been the subject of many legal decisions. It is, however, very difficult for the Contractor to succeed where his claims for disruption are based simply on global overspent on its resources and he normally has to isolate the cause of disruption and evaluate its effect.[2] However, there is a recent tendency for claims consultants to adopt the "measured mile" approach, which compares the production of a relatively undisrupted period (or in the absence of an undisrupted period, the production based on originally planned resources and timing) with the production of a period in which there are a lot of late instructions and variation orders, and to claim for extra management charges and disruption costs. Amongst the Government and the consultants, there have also been a lot of discussions concerning valuation under Clause 61 or under Clause 63(b) for any other "knock on" effect of the variation. Valuation under Clause 61 can include profit while valuation under Clause 63(b) allows for Cost only.[3]

Analysis and application

Clause 63 "… the Contractor has been or is likely to be involved in expenditure for which the Contractor would not be reimbursed by a payment made under any other provision in the Contract by reason of the progress of the Works … having been materially affected by: …"

For this Clause to apply, the disruptive event must affect the progress of the Works although not necessarily affect the critical path as discussed in the commentary. The wording "would not be reimbursed by a payment made under any other provision" suggests that the direct Cost of the subject events (a) to (e) may have already been valued separately. For example, the labour and plant cost of opening up for inspection has already been valued and the disruptive effect such as stoppages of other activities would be valued under this clause. In many cases, the entitlement under "other clauses" and this Clause overlaps, e.g., valuation of variation under Clause 61 and valuation under Clause 63(b). However, any suggestion to value prolongation cost under Clause 61 instead of Clause 63(b) should be treated with caution. For a detailed discussion, see commentary to the proviso to Clause 61(1).

Clause 63(a) "the Contractor not having received in due time necessary instruction, orders, directions, decisions, Drawings, Specification … for which the Contractor specifically applied in writing on a date which, having regard to the time for completion … was neither unreasonably distant from nor unreasonably close to the date on which it was necessary for the Contractor to receive the same,"

By this Clause, the Contractor must request the information at the right time, allowing the Engineer adequate time to make the information available. This Clause is complementary to Clause 6(4), under which the Contractor shall give adequate notice in writing to the Engineer regarding other Drawings or Specification that may be required for the execution of the Works. As pointed out in the commentaries to Clauses 5 and 6, many contractors make systematic requests for information, direction and clarification, creating a lot of pressure and extra workload for the Engineer and also forming a basis for a huge claim in future.

Clause 63(b) "any variation ordered in accordance with Clause 60, …"

The implication has been explained in the commentary.

Clause 63(c) "the opening up for inspection … any work covered up or the testing of materials or workmanship not required by the Contract but directed by the Engineer … unless the inspection or test showed that the work, material or workmanship were not in accordance with the Contract, …"

This sub-clause (c) allows compensation to the Contractor for any disruptive effects of additional inspection or test required by the Engineer or the Engineer's Representative unless such inspection or test reveals unsatisfactory work, material or workmanship.

Clause 63(d) "delay caused by any person or any company, not being a utility undertaking, engaged by the Employer in supplying materials or in executing work directly connected with but not forming part of the Works, …"

If such works by the Employer's own contractor or supplier are required and although such works are described in the Specification, it is advisable to describe explicitly that the disruptive effects of such works shall be allowed for by the Contractor as part of his general obligation or otherwise are paid for under separate items. This would avoid a claim under this sub-clause.

Clause 63(e) "late delivery of material, Plant or equipment by the Employer, …"

This sub-clause is straightforward and self-explanatory.

Comparison with the ICE Conditions

It is recognized that in the ICE Conditions, there is no single clause similar to the HKSAR Clause 63 summarizing the different events which entitle the Contractor to Cost reimbursement arising from their disruptive effect to the progress of the Works. The ICE sub-clause 7(4) provides for extension of time and/or reimbursement as a result of late supply of Drawings, Specification or giving of instructions requested by the Contractor. The ICE sub-clauses 27(2b), (3) and (4) provide for extension of time and/or reimbursement arising from the requirement of the Contractor having to comply with additional conditions or limitation imposed as a result of the Employer obtaining any street work licence. Clause 38(2) provides for extension of time and/or reimbursement arising from the Contractor having to uncover any part of the Works or making opening for examination as a result of the Engineer's direction which are later found to be unnecessary. For variation order or late delivery of materials, plant or equipment by the Employer, the disruptive effect could generally be covered by valuation of variation under Clause 52.

Comparison with the FIDIC Conditions

For the FIDIC Conditions, there is also no single clause similar to the HKSAR Clause 63. Sub-clauses 1.9, 8.5 and 11.8 have similar effect to the ICE sub-clauses 7(4), 27(2)(b), (3) and (4), and 38(2) respectively. However, the FIDIC Clause 8.5 has a much wider coverage than the ICE Clause 27 or the HKSAR Clause 63(b) in that the Contractor is entitled to extension of time and Cost reimbursement arising from the disruptive effect of procedures laid down by the relevant legally constituted public authorities in the country provided that the Contractor has diligently followed such procedures and yet the delay and disruption were unforeseeable.

Notes

1. *John Doyle Construction Ltd. v Laing Management (Scotland) Ltd.* [2004] ABC, LR06/11.
2. For methods of evaluating disruption, see Roger Knowles, *150 Contractual Problems and Their Solutions*, Section 9.15.
3. The *Jesse Grove Report* (1998) says "There is an overlap between Clause 63(b) and the proviso of Clause 61 in that both provide for additional payment in respect of impact on unchanged work by reason of a variation order (the "ripple" or "knock-on" effect), but the former contemplates Cost recovery while the latter allows for more general price adjustment." See also *Government Practice Note on GCC 61 and 63(b)*.

 Q 38 In order to rectify some design deficiencies, the Engineer instructed that several precast segments have to be cast 200 mm shorter than the original design. What provision can the Contractor rely on in recovering delay and additional expenses?

Clause 64

Notice of Claims

(1) If the Contractor intends to claim a higher rate than one notified to him by the Engineer pursuant to Clause 59(4)(b) or Clause 61(2) or Clause 84(4)(b) the Contractor shall within 28 days of such notification give notice in writing of his intention to claim to the Engineer.

(2) If the Contractor intends to claim any additional payment under the provisions of any Clause of the General Conditions of Contract or Special Conditions of Contract other than as mentioned in sub-clause (1) of this Clause, the Contractor shall within 28 days after the happening of the events giving rise to a claim serve notice in writing on the Engineer of his intention to claim and the contractual provisions upon which the claim is based.

(3) The Contractor shall keep such contemporary records as may reasonably be necessary to support any claim and shall give to the Engineer details of the records being kept in respect thereof. Without necessarily admitting the Employer's liability, the Engineer may require the Contractor to keep and agree with the Engineer's Representative any additional contemporary records as are reasonable and may in the opinion of the Engineer be material to the claim. The Contractor shall permit the Engineer and the Engineer's Representative to inspect all records kept pursuant to this Clause and shall supply copies thereof as and when the Engineer or Engineer's Representative shall so require.

Corresponding clauses in the ICE Conditions:

Clause 53

(1) If the Contractor intends to claim a higher rate or price than one notified to him by the Engineer pursuant to sub-clauses (3) and (4) of Clause 52 or Clause 56(2) the Contractor shall within 28 days after such notification give notice in writing of his intention to the Engineer.

(2) If the Contractor intends to claim any additional payment pursuant to any Clause of these Conditions other than sub-clauses (3) and (4) of Clause 52 or Clause 56(2) he shall give notice in writing of his intention to the Engineer as soon as may be reasonable and in any event within 28 days after the happening of the events giving rise to the claim.

(3) Without necessarily admitting the Employer's liability the Engineer may upon receipt of a notice under this Clause instruct the Contractor to keep such contemporary records or further contemporary records as the case may be as are reasonable and may be material to the claim of which notice has been given and the Contractor shall keep such records.

The Contractor shall permit the Engineer to inspect all records kept pursuant to Clause 53 and shall supply him with copies thereof as and when the Engineer shall so instruct.

Corresponding clauses in the FIDIC Conditions:

Clause 20.1

If the Contractor considers himself to be entitled to any extension of the Time for Completion and/or any additional payment, under any Clause of these Conditions or otherwise in connection with the Contract, the Contractor shall give notice to the Engineer, describing the event or circumstance giving rise to the claim. The notice shall be given as soon as practicable, and not later than 28 days after the Contractor became aware, or shall have become aware, of the event or circumstance.

If the Contractor fails to give notice of a claim within such period of 28 days, the Time for Completion shall not be extended, the Contractor shall not be entitled to additional payment, and the Employer shall be discharge from all liability in connection with the claim. Otherwise, the following provisions of this Sub-Clause shall apply.

The Contractor shall also submit any other notices which are required by the Contract, and supporting particulars for the claim, all as relevant to such event or circumstance.

The Contractor shall keep such contemporary records as may be necessary to substantiate any claim, either on the Site or at another location acceptable to the Engineer. Without admitting the Employer's liability, the Engineer may, after receiving any notice under this Sub-Clause, monitor the record-keeping and/or instruct the Contractor to keep further contemporary records. The Contractor shall permit the Engineer to inspect all these records, and shall (if instructed) submit copies to the Engineer.

條款 64

香港特別行政區土木工程合同一般條件：

索償通知

(1) 如「承建商」有意索償比「工程師」根據第59條(4)款(b)或第61條(2)款或第84條(4)款(b)通知為高的價率，「承建商」須在「工程師」發給通知後28天內以書面通知「工程師」其索償意向。

(2) 如「承建商」有意在本條第(1)款提及之外依據「合同一般條件」或「合同特別條件」內任何條文索償任何額外付款，「承建商」必須在導致索償事件發生後28天內以書面通知「工程師」其索償意向及說明該索償所依據的「合同」條文。

(3) 「承建商」必須保有為支持任何索償而有合理需要的當時紀錄，並須向「工程師」提交這些紀錄的細節。「工程師」在無須承認「僱主」有法律責任的情況下，可要求「承建商」保有得「工程師代表」同意的任何額外的當時紀錄，而此等紀錄是合理而且「工程師」認為對索償是重要的。「承建商」必須允許「工程師」及「工程師代表」查閱按照本條去保有的一切紀錄，並須在「工程師」及「工程師代表」要求時向他們提供這些紀錄的副本。

Commentary

This Clause 64 stipulates the required procedures to be followed by the Contractor in submitting a claim for a higher rate than the one notified to him by the Engineer after his valuation or for any additional payment under the provision of any Clause of the Conditions of Contract. There are two essential distinct steps in claim submission, namely, the giving of notice in writing and then as soon as reasonable the sending to the Engineer interim accounts giving full and detailed particulars of the claim, including all necessary calculations. Failure to comply with the first step regarding notice requirement shall likely invalidate the Contractor's claim because the HKSAR Conditions treat the notice requirement as a condition precedent[1] pursuant to sub-clause (5). Failure to submit detailed particulars and proper interim accounts is, however, considered to have only prejudicial effects pursuant to sub-clause (6). Apart from giving notice of claim in a diplomatic manner, the Contractor has no difficulty in complying with this condition precedent. Notice requirement, from the Employer's perspective, helps to limit surprise arising from late or unmeritorious claims and provide a means of effective financial control. There is distinct merit in retaining it as a condition precedent despite of a contrary recommendation from an international expert.[2] On the other hand, the Contractor's attitude in complying with timely submission of full and detailed particulars and proper interim accounts is usually less enthusiastic and some contractor prefer to adopt a wait and see attitude. The Engineer has difficulty in compelling the Contractor to do so if the latter does not co-operate. The Contractor may save all his efforts in future negotiation, mediation or in arbitration and in many cases, his claims are not effectively prejudiced. Despite this, as a good management practice, the Engineer is well advised to keep his own detailed records, write to the Contractor to request compliance and give suitable warning about any prejudicial effects on the claim. If the Contractor consistently breaches his obligation, the Employer may reflect such deficiencies in the performance report or even consider recovering from the Contractor the expenses of preparing such records.[3]

Analysis and application

Clause 64 "(1) If the Contractor intends to claim a higher rate than one notified to him by the Engineer … the Contractor shall within 28 days … give notice in writing … (2) If the Contractor intends to claim any additional payment under the provision of any clause … the Contractor shall within 28 days after the happening of the events giving rise to a claim serve notice in writing …"

As discussed before, the essential points in these sub-clauses are whether the notice requirement constitutes a condition precedent and previous legal cases confirm it is. It is also submitted that this Clause 64 is not invalidated by the provisions of the *Control of Exemption Clause Ordinance*[4] since Clause 64 itself is considered reasonable. The wording "to claim a higher rate than one notified to him by the Engineer" may suggest that if the rate has not yet been formally fixed and notified to the Contractor probably because the Engineer cannot foresee any disagreement and the Contractor has not raised any dispute on any provisional rate, the Contractor is not bound to give notice to claim a varied rate yet. This may create future uncertainties. The wording "to claim any additional payment under the provision of any Clause" suggests that any claim for breach of Contract, such as the breach of an implied term that the Engineer will act in a reasonable manner, or outside the Contract, such as a claim for acceleration,[5] is not caught by the notice requirement. Regarding the notice itself, there is no particular format and content requirements and the mere mentioning of the Contractor's intention to make the particular claim will suffice. It is to be noted that even the claim is based on an incorrect provision in the notice which is subsequently amended to a correct provision in the Conditions of Contract will not invalidate nor prejudice the Contractor's final claim.[6]

Clause 64(3) "The Contractor shall keep such contemporary records as may reasonably be necessary … and shall give to the Engineer details of the records … Without necessarily admitting the Employer's liability, the Engineer may require the Contractor to keep and agree with the Engineer's Representative any additional contemporary records as are reasonable … The Contractor shall permit the Engineer and the Engineer's Representative to inspect all records … and shall supply copies thereof …"

For payment claims, the keeping of records which are "contemporary" is very important. The Court held that a sufficient connection between that which is to be recorded and the act of recording was required. Whether a record is contemporary or not will depend on the fact surrounding the making of that record, with due regard to the custom and the practice of the industry and that it would be exceptional if a document or statement (including the witness statement) could be regarded as contemporary if made more than a few weeks after the event it records.[7] With regard to giving of records to the Engineer, the duty of the Contractor is automatic and not dependent upon the request from the Engineer. In a potential claims situation, the Engineer is advised to invoke the provision of the second sentence to request the Contractor to keep additional records and agree on those with the Engineer's Representative. In major claims, the Engineer should be proactive enough to inspect records kept by the Contractor and request him to supply copies of such records where necessary.

 Q 39 An additional water-cooling main is required to be laid under the suspended floor. What procedures should the Contractor follow to recover time and additional expenses?

 The school adjacent to the uncompleted flyover is now being utilized to host international examination every Thursday afternoon. The Education Bureau does not want the construction noise to disturb the students. What are the Contractor's entitlements? What procedures should the Contractor follow?

(4) After the giving of a notice to the Engineer under this Clause, the Contractor shall, as soon as is reasonable, send to the Engineer a first interim account giving full and detailed particulars of the circumstances giving rise to the claim, the rate or sum claimed and the manner in which such rate or sum is calculated. Thereafter, at such intervals as the Engineer may reasonably require, the Contractor shall send to the Engineer further up-to-date accounts giving the accumulated total of the claim and any further full and detailed particulars in relation thereto.

(5) If the Contractor fails to comply with the notice provisions contained in sub-clauses (1) or (2) of this Clause in respect of any claim, such claim shall not be considered.

(6) If the Contractor fails to comply with the provisions of sub-clauses (3) or (4) of this Clause in respect of any claim the Engineer may consider such claim only to the extent that the Engineer is able on the information made available.

Provided that the Engineer shall not be obliged to take into account when considering a claim any particulars of the claim received by him after the expiry of a period of 180 days calculated from the date of completion stated in the certificate of completion with respect to the Works. In the event of different certificates of completion having been issued for different Sections or parts of the Works pursuant to Clause 53, the expression "certificate of completion" shall, for the purpose of this sub-clause, mean the last of such certificates.

Corresponding clauses in the ICE Conditions:

Clause 53 (*continued*)

(4) After the giving of a notice to the Engineer under this Clause the Contractor shall as soon as is reasonable in all the circumstances send to the Engineer a first interim account giving full and detailed particulars of the amount claimed to that date and of the grounds upon which the claim is based.

Thereafter at such intervals as the Engineer may reasonably require the Contractor shall send to the Engineer further up to date accounts giving the accumulated total of the claim and any further grounds upon which it is based.

(5) If the Contractor fails to comply with any of the provisions of this Clause in respect of any claim which he shall seek to make then the Contractor shall be entitled to payment in respect thereof only to the extent that the Engineer has not been prevented from or substantially prejudiced by such failure in investigating the said claim.

(6) The Contractor shall be entitled to have included in any interim payment certified by the Engineer pursuant to Clause 60 such amount in respect of any claim as the Engineer may consider due to the Contractor provided that the Contractor shall have supplied sufficient particulars to enable the Engineer to determine the amount due.

If such particulars are insufficient to substantiate the whole of the claim the Contractor shall be entitled to payment in respect of such part of the claim as the particulars may substantiate to the satisfaction of the Engineer.

Corresponding clauses in the FIDIC Conditions:

Clause 20.1 (*continued*)

Within 42 days after the Contractor became aware (or should have become aware) of the event or circumstance giving rise to the claim, or within such other period as may be proposed by the Contractor and approved by the Engineer, the Contractor shall send to the Engineer a fully detailed claim which includes full supporting particulars of the basis of the claim and of the extension of time and/or additional payment claimed. If the event or circumstance giving rise to the claim has a continuing effect:

(a) this fully detailed claim shall be considered as interim;

(b) the Contractor shall send further interim claims at monthly intervals, giving the accumulated delay and/or amount claimed, and such further particulars as the Engineer may reasonably require; and

(c) the Contractor shall send a final claim within 28 days after the end of the effects resulting from the event or circumstance, or within such other period as may be proposed by the Contractor and approved by the Engineer.

Within 42 days after receiving a claim or any further particulars supporting a previous claim, or within such other period as may be proposed by the Engineer and approved by the Contractor, the Engineer shall respond with approval, or with disapproval and detailed comments. He may also request any necessary further particulars, but shall nevertheless give his response on the principles of the claim within such time.

Each Payment Certificate shall include such amounts for any claim as have been reasonably substantiated as due under the relevant provision of the Contract. Unless and until the particulars supplied are sufficient to substantiate the whole of the claim, the Contractor shall only be entitled to payment for such part of the claim as he has been able to substantiate.

The Engineer shall proceed in accordance with Sub-Clause 3.5 [*Determinations*] to agree or determine (i) the extension (if any) of the Time for Completion (before or after its expiry) in accordance with Sub-Clause 8.4 [*Extension of Time for Completion*], and/or (ii) the additional payment (if any) to which the Contractor is entitled under the Contract.

The requirements of this Sub-Clause are in addition to those of any other Sub-Clause which may apply to a claim. If the Contractor fails to comply with this or another Sub-Clause in relation to any claim, any extension of time and/or additional payment shall take account of the extent (if any) to which the failure has prevented or prejudiced proper investigation of the claim, unless the claim is excluded under the second paragraph of this Sub-Clause.

條款 64	香港特別行政區土木工程合同一般條件：

索償通知

（續）

(4) 根據本條規定，在給予「工程師」一份通知後，「承建商」必須在合理時間內盡快交給「工程師」第一份中期帳目，完全及詳盡地記載導致索償的情況、價率或金額，以及計算此等價率或金額的方法。然後，「承建商」必須在每隔一段由「工程師」合理地要求的時間內交給他進一步的最新帳目，記載索償的累積總額及任何與索償有關的進一步的完全和仔細的詳情。

(5) 如「承建商」沒有按照本條第(1)或(2)款所載述關於通知的條文提出索償，此項索償不會獲考慮。

(6) 如「承建商」沒有按照本條第(3)或(4)款的條文提出索償，「工程師」只會根據他所得的資料在所能的範圍內考慮該項索償。

但是，「工程師」在考慮任何索償時，無責任考慮在「本工程」竣工證明書內所述的竣工日期翌日起計的180天後才接獲的任何與索償有關的情況。如根據第53條就「本工程」的不同工段或部分已頒發不同的竣工證明書，則對本款而言，「竣工證明書」 是指此等證明書最後的一張。

Analysis and application

Clause 64(4) "… as soon as is reasonable, send to the Engineer a first interim account giving full and detailed particulars of the circumstances giving rise to the claim, …"

As discussed in the commentary, there is not much the Engineer can do if the Contractor refuses to comply with furnishing full and detailed particulars of his claim. Drastic means such as withholding certificates will be impracticable. The assessment of extension of time and the associated prolongation cost, which is usually where the large sum of money lies, depends on the Contractor submitting realistic programmes and the detailed cause and effect of the relevant events. It is recommended that the Engineer must not sit back and wait but should take a proactive attitude to seek and obtain such information. Very often the Engineer needs to rectify the defects in the Contractor's programme and update the progress to facilitate his assessment of the extension of time and hence the disruptive and prolongation costs.

Clause 64(5) "If the Contractor fails to comply with the notice provisions contained in sub-clauses (1) or (2) of this Clause in respect of any claim, such claim shall not be considered."

This Clause states that the notice requirement is a condition precedent, as discussed earlier, and the Contractor's claim will be barred if he fails to give notice of claim within the specified time limit.

Clause 64(6) "If the Contractor fails to comply with the provisions of sub-clauses (3) or (4) of this Clause … the Engineer may consider such claim only to the extent that the Engineer is able on the information made available."

This sub-clause, if effectively used, enables the Engineer to make an assessment as far as possible based on the information available to him, partly submitted by the Contractor and partly through his own Site records. It is a good practice for the Engineer to review all information at the relevant time and make an assessment rather than leave it in the hands of the Contractor who may leave everything till the end of the Contract and come up with numerous claims in the hope of getting an overall settlement with the Employer based on a large amount of information which is not contemporary. The Employer may be exposed not only to a huge claim but also to heavier financial charges.

Clause 64 "… Provided that the Engineer shall not be obliged to take into account when considering a claim any particulars of the claim received by him after the expiry of a period of 180 days calculated from the date of completion …"

As stated in the commentary to sub-clause (3), the availability of contemporary records is very important to support the merit and the quantum of a claim. The above proviso gives an absolute bar to any new information to support the claim approximately 6 months after the date of completion. If landscape establishment works are included in the Contract, the last certificate is usually more than 12 months after completion of the major civil engineering works.

Comparison with the ICE Conditions

This ICE Clause 53 is very similar to the HKSAR Clause 64. Although it does not say that the Contractor shall keep contemporary records automatically as required by the HKSAR Clause, the requirement in the ICE Clause that the Contractor shall submit a first account and then up-to-date accounts giving full and detailed particulars of the amount claimed implicitly obligates the Contractor to keep contemporary records to substantiate his claims. The ICE Clause does not give an absolute bar to submission of new information on the claim whereas the HKSAR Clause gives an absolute bar 180 days after completion of the Works.

Comparison with the FIDIC Conditions

The FIDIC Clause makes it a condition precedent of the notice requirement for both monetary and extension of time claims. However, this relatively new provision has not been substantially tested, especially with claim for extension of time. There is a danger that time could be set at large if an act of prevention by the Employer causes delay[8] even though the Contractor fails to submit the required notice within the specified time barring his claim for extension of time. The FIDIC Clause also allows for the situation where the claim has a continuing effect, e.g. a variation order on the pile design affecting many structures. Another difference is the setting of a time limit for the Contractor to submit his claim details within 42 days after he became aware of the event or circumstances giving rise to the claim and also 42 days for the Engineer to give approval or disapproval and to give his response on the principles of the claim after receiving it.

Notes

1. *Blackford & Sons (Calne) Ltd. v Christchurch Borough* [1962] I Lloyd's Rep. 349; *Chiemgauer Membran Und Zeltban GmbH v New Millennium Experience Company Ltd.* CA (Civil Division) (1999).
2. *Recommendations in Grove Report on General Conditions of Contract.*
3. Max Abrahamson, *Engineering Law and the ICE Contracts*, 4th Edition, p. 194, n. 4.
4. *Control of Exemption Clause Ordinance* (Cap 71).

5.   Max Abrahamson, *Engineering Law and the ICE Contracts*, 4th Edition, p. 194, n. 3.

6.   Max Abrahamson, *Engineering Law and the ICE Contracts*, 4th Edition, p. 194, n. 6.

7.   *Attorney General for the Falkland Islands v Gordon Forbes Construction (Falklands) Ltd.* [2003] BLR280.

8.   *Gaymark Investment Pty. Ltd. v Walter Construction Group Ltd.* [1999] NTSC143.

 Q 41 The Engineer is to review the details of reinforcement in the cross-beams. It is unlikely that work can be resumed within one week. What should the Contractor do? List out the actions step by step.

Clause 68

Accounting of Provisional and Contingency Sums

Provisional Sums and the Contingency Sum shall be deducted from the Contract Sum and in lieu thereof shall be added the value of the work ordered by the Engineer, valued in accordance with Clause 61 or, where the Provisional Sum or Contingency Sum relates to work executed by, or goods, materials or services supplied by a Nominated Sub-contractor, valued in accordance with the sub-contract, deducting any trade or other discount. Any part of any Provisional Sum or the Contingency Sum which is to be used for work to be executed by, or goods, materials or services to be supplied by a Nominated Sub-contractor, shall be deemed to be a Prime Cost Sum.

Corresponding clauses in the ICE Conditions:

Clause 58

(1) In respect of every Provisional Sum the Engineer may order either or both of the following.

(a) Work to be carried out or goods materials or services to be supplied by the Contractor the value thereof being determined in accordance with Clause 52 and included in the Contract Price.

(b) Work to be carried out or goods materials or services to be supplied by a Nominated Sub-contractor in accordance with Clause 59.

(2) In respect of every Prime Cost Item the Engineer may order either or both of the following.

(a) Subject to Clause 59 that the Contractor employ a sub-contractor nominated by the Engineer for the carrying out of any work or the supply of any goods materials or services included therein.

(b) With the consent of the Contractor that the Contractor himself carry out any such work or supply any such goods materials or services in which event the Contractor shall be paid in accordance with the terms of a quotation submitted by him and accepted by the Engineer or in the absence thereof the value shall be determined in accordance with Clause 52 and included in the Contract Price.

(3) If in connection with any Provisional Sum or Prime Cost Item the services to be provided include any matter of design or specification of any part of the Permanent Works or of any equipment or plant to be incorporated therein such requirement shall be expressly stated in the Contract and shall be included in any Nominated Sub-contract. The obligation of the Contractor in respect thereof shall be only that which has been expressly stated in accordance with this sub-clause.

Clause 59

(5) For all work carried out or goods materials or services supplied by Nominated Sub-contractors there shall be included in the Contract Price

(a) the actual price paid or due to be paid by the Contractor in accordance with the terms of the sub-contract (unless and to the extent that any such payment is the result of a default of the Contractor) net of all trade and other discounts, rebates and allowances other than any discount obtained by the Contractor for prompt payment.

Corresponding clauses in the FIDIC Conditions:

Clause 13.5

Each Provisional Sum shall only be used, in whole or in part, in accordance with the Engineer's instruction, and the Contract Price shall be adjusted accordingly. The total sum paid to the Contractor shall include only such amounts, for the work, supplies or services to which the Provisional Sum relates, as the Engineer shall have instructed. For each Provisional Sum, the Engineer may instruct:

(a) work to be executed (including Plant, Materials or services to be supplied) by the Contractor and valued under Sub-Clause 13.3 [*Variation Procedure*]; and/or

(b) Plant, materials or services to be purchased by the Contractor, from a nominated subcontractor (as defined in Clause 5 [*Nominated Subcontractors*]) or otherwise; and for which there shall be included in the Contract Price:

(i) the actual amounts paid (or due to be paid) by the Contractor, and

(ii) a sum for overhead charges and profit, calculated as a percentage of these actual amounts by applying the relevant percentage rate (if any) stated in the appropriate Schedule. If there is no such rate, the percentage rate stated in the Appendix to Tender shall be applied.

The Contractor shall, when required by the Engineer, produce quotations, invoices, vouchers and accounts or receipts in substantiation.

條款 68

香港特別行政區土木工程合同一般條件:

「備用金額」及「應變金額」的計算

「備用金額」及「應變金額」必須從「合同金額」中扣除,取而代之加入「工程師」命令實施的工程價值,並必須根據第61條來釐訂其價值;或當「備用金額」或「應變金額」是關於「指定次承判商」所實施的工程或供應的貨物、物料或提供的服務時,工程價值必須根據分包合同來釐訂,並須扣除任何商業或其他折扣。「備用金額」或「應變金額」內如有任何部分是用來支付由「指定次承判商」所實施的工程或供應的貨物、物料或提供的服務,此等部分會被視為「基本成本款項」。

Commentary

This Clause 68 is seldom referred to by the Engineer partly because the Contract is a re-measurement Contract in which payment is made on the measurement of the quantity of works actually carried out in accordance with the Contract, and partly because the Government of the HKSAR now seldom specifies Nominated Sub-contractor in the Contract. However, the similar clause in Design and Build Contract is much more important because the Contract is paid on a lump sum basis. The use of a Nominated Sub-contractor is seldom the best arrangement mainly because there is no contractual bond between the Employer and the Nominated Sub-contractor and in case there is any delay, the Contractor is entitled to an extension of time under Clause 50(1)(b)(x), but the Employer is not able to recover damages from the Nominated Sub-contractor. Apart from this, there are many other problems such as the responsibility for the choice of materials, design, defective work, and non-performance, etc. in the use of Nominated Sub-contractors.[1] To overcome the difficulties, the Government of the HKSAR keeps a List of Specialist Contractors and Suppliers and a Special Condition requires that the Contractor enters into sub-contracts with the contractors or suppliers in the said List for the carrying out of specialist works or supply of specialist materials, such as bitumen surfacing works, bridge bearings or expansion joints. This has the advantage of availability of expertise in the construction of specialist works and yet all the responsibilities fall squarely on the Contractor.

Analysis and application

Clause 68 "Provisional Sums and the Contingency Sum shall be deducted from the Contract Sum and in lieu thereof shall be added the value of the Work ordered by the Engineer, valued in accordance with Clause 61 ..."

The Provisional Sum is defined as a sum provided for work or expenditure which has not been quantified or detailed at the time the Tender documents are issued, which sum may include provision for work to be executed or for materials or services supplied by a Nominated Sub-contractor. It should be distinguished from the "Provisional Item" which is defined in the Method of Measurement as an item with quantity and rate and which will only be carried out upon the instruction of the Engineer. Though the Provisional Sum can only be expended also upon the instruction of the Engineer, unlike the Provisional Item, a variation order is required and the works have to be valued in accordance with Clause 61.

Clause 68 "… where the Provisional Sum or Contingency Sum relates to work … by a Nominated Sub-contractor, valued in accordance with the sub-contract, deducting any trade or other discount … goods, materials or services to be supplied by a Nominated Sub-contractor, shall be deemed to be a Prime Cost Sum."

The Prime Cost Sum is defined in Clause 1(1) as "the sum provided for work to be executed or for materials or services to be supplied by a Nominated Sub-contractor, such sum shall be the estimated net price." For the Prime Cost Sum, the wording "to be executed or for materials or services to be supplied" suggests that it will be used by the Contractor who is entitled to the sum for attendance under Clause 67(2) if it is omitted. Also, unlike the Provisional Sum, the Contractor may refuse to carry out the work under a Prime Cost Sum himself if such work is not carried out by the Nominated Sub-contractor.[2] In any case, as stated in the commentary, the Prime Cost Sum is seldom used nowadays in civil engineering contracts in Hong Kong.

Comparison with the ICE Conditions

The principle for the use and valuation of work carried out under the Provisional Sum or Prime Cost Item (despite its being called Prime Cost Item instead of Prime Cost Sum) are essentially the same.

Comparison with the FIDIC Conditions

The FIDIC Conditions do not have a definition for the Prime Cost Sum or Prime Cost Item, which, it is submitted, does not serve any useful purpose. This is because the Provisional Sum, for all intents and purposes, covers the work, plant, materials or services to be supplied by the Nominated Sub-contractor.

Notes

1. Max Abrahamson, *Engineering Law and the ICE Contracts*, 4th Edition, pp. 224–234.
2. Max Abrahamson, *Engineering Law and the ICE Contracts*, 4th Edition, p. 235, n. 1.

Q 42   The Employer suddenly wishes to produce a short video for the construction of the viaduct. Should he use Provisional Sum or Contingency Sum?

Clause 71

The HKSAR Conditions:

Vesting of Constructional Plant and Temporary Buildings

All Constructional Plant and temporary buildings owned by the Contractor shall when brought onto the Site be and become the property of the Employer but may be removed from the Site by the Contractor at any time unless removal is expressly prohibited by the Engineer in writing. Upon removal as aforesaid or under the terms of Clause 88(2) such Constructional Plant and temporary buildings shall re-vest in the Contractor. Upon completion of the Works the remainder of such Constructional Plant and temporary buildings shall, subject to Clause 81, re-vest in the Contractor.

Clause 72

The HKSAR Conditions:

Vesting of Materials

All materials owned by the Contractor for incorporation in the Works shall be and become the property of the Employer upon delivery to the Site, and shall not be removed without an instruction or the prior written consent of the Engineer. Materials shall, subject to Clause 81, only re-vest in the Contractor to the extent that they may be found to be surplus to requirements upon or prior to completion of the Works. The operation of this Clause shall not be deemed to imply any approval by the Engineer of such materials or prevent the rejection by the Engineer of any material at any time.

Corresponding clauses in the ICE Conditions:

Clause 54

(1) No Contractor's Equipment Temporary Works materials for Temporary Works or other goods or materials owned by the Contractor and brought on to the Site for the purposes of the Contract shall be removed without the written consent of the Engineer which consent shall not unreasonably be withheld.

(2) (*Not quoted*)

(3) (*Not quoted*)

(4) With a view to securing payment under Clause 60(1)(c) the Contractor may (and shall if the Engineer so directs) transfer to the Employer the property in goods and materials listed in the Appendix to the Form of Tender or as subsequently agreed between the Contractor and the Employer before the same are delivered to the Site provided that the goods and materials

(a) have been manufactured or prepared and are substantially ready for incorporation in the Works and

(b) are the property of the Contractor or the contract for the supply of the same expressly provides that the property therein shall pass unconditionally to the Contractor upon the Contractor taking the action referred to in sub-clause (5) of this Clause and

(c) have been marked and set aside in accordance with sub-clause (5) of this Clause.

(5) The intention of the Contractor to transfer the property in any goods or materials to the Employer in accordance with this Clause shall be evidenced by the Contractor taking or causing the supplier of those goods or materials to take the following action.

(a) Provide to the Engineer documentary evidence that the property of the said goods or materials has vested in the Contractor.

(b) Suitably mark or otherwise plainly identify the goods and materials so as to show that their destination is the Site that they are the property of the Employer and (where they are not stored at the premises of the Contractor) to whose order they are held.

(c) Set aside and store the said goods and materials so marked and identified to the satisfaction of the Engineer.

(d) Send to the Engineer a schedule listing and giving the value of every item of the goods and materials so set aside and stored and inviting him to inspect them.

(6) Upon the Engineer approving in writing the transfer in ownership of any goods and materials for the purposes of this Clause they shall vest in and become

the absolute property of the Employer and thereafter shall be in the possession of the Contractor for the sole purpose of delivering them to the Employer and in incorporating them in the Works and shall not be within the ownership control or disposition of the Contractor. Provided always that

(a) approval by the Engineer for the purposes of this Clause or any payment certified by him in respect of goods and materials pursuant to Clause 60 shall be without prejudice to the exercise of any power of the Engineer contained in this Contract to reject any goods or materials which are not in accordance with the provisions of the Contract and upon any such rejection the property in the rejected goods or materials shall immediately re-vest in the Contractor and

(b) the Contractor shall be responsible for any loss or damage to such goods or materials and for the cost of storing handling and transporting the same and shall effect such additional insurance as may be necessary to cover the risk of such loss or damage from any cause.

(7) (*Not quoted*)

(8) (*Not quoted*)

(9) (*Not quoted*)

Corresponding clauses in the FIDIC Conditions:

Clause 4.17

The Contractor shall be responsible for all Contractor's Equipment. When brought on to the Site, Contractor's Equipment shall be deemed to be exclusively intended for the execution of the Works. The Contractor shall not remove from the Site any major items of Contractor's Equipment without the consent of the Engineer. However, consent shall not be required for vehicles transporting Goods or Contractor's Personnel off Site.

(*The following paragraph is optional*)
Contractor's Equipment which is owned by the Contractor (either directly or indirectly) shall be deemed to be the property of the Employer with effect from its arrival on the Site. This vesting of property shall not:

(a) affect the responsibility or liability of the Employer,

(b) prejudice the right of the Contractor to the sole use of the vested Contractor's Equipment for the purpose of the Works, or

(c) affect the Contractor's responsibility to operate and maintain Contractor's Equipment.

The property in each item shall be deemed to revest in the Contractor when he is entitled either to remove it from the Site or to receive the Taking-Over Certificate for the Works, whichever occurs first.

條款 71

香港特別行政區土木工程合同一般條件：

「施工設備」及「臨時建築物」的擁有權

「承建商」所擁有的所有「施工設備」及「臨時建築物」，一經携帶入工地，即為並成為「僱主」的財產。除非「工程師」以書面明文地禁止移走，否則這些「施工設備」及「臨時建築物」可由「承建商」隨時移離「工地」。但若如上述情況或根據第88條(2)款一旦移出「工地」，此等「施工設備」及「臨時建築物」即重新歸「承建商」擁有。在「本工程」竣工時，除第81條另有規定外，餘下的此等「施工設備」及「臨時建築物」即重新歸「承建商」擁有。

條款 72

香港特別行政區土木工程合同一般條件：

物料的擁有權

「承建商」所擁有而會納入「本工程」的所有物料一經送到「工地」，即為並成為「僱主」的財產，未有「工程師」指令或事前以書面同意，不可移離「工地」。除第81條另有規定外，只有在「本工程」竣工時或竣工前所多於需要的剩餘物料才可重新歸「承建商」擁有。本條的實行不可被視作隱含此等物料已獲「工程師」批准或阻止「工程師」在任何時間拒絕接受任何物料。

Commentary

Clause 71 on the vesting of Constructional Plant and temporary buildings has been in the HKSAR Conditions for several decades but has not been very useful, partly because it only applies to the Plant owned by the Contractor (excluding hired Plant or hire-purchase Plant) and partly because there is no practical need to retain the Contractor's Plant unless the Plant is for work of a specialized nature. Further, the vesting provision is practically seldom followed through by the Engineer, because it is not seen by the Engineer to be of any significance unless forfeiture is contemplated. Even such course of action is finally adopted, or the Contractor is bankrupt or under liquidation proceedings, the Plant will be removed in the earliest instance by the Contractor before the Employer is ready to take any action. In 2004, a Special Condition became mandatory and this Special Condition requires those Plants essential for the completion of the Works to be subject to proof of ownership, and if the Plant is not owned by the Contractor, an undertaking from the Plant owner is needed to assign to the Employer the benefit of any hiring or hire-purchase agreement made with the Contractor. In case of termination of Contract by the Employer or the abandonment of Contract by the Contractor, the Employer can then use those essential Plant for the completion of the Works.

Clause 72 regarding vesting of materials is relatively straightforward since most materials are to be incorporated into the Works and the Contractor would have been paid for the materials once such materials arrive at the Site.

Analysis and application

Clause 71 "All Constructional Plant and temporary buildings owned by the Contractor shall when brought onto the Site be and become the property of the Employer ..."

By definition 1(1), "Constructional Plant" means "all appliances or things of whatsoever nature required for the execution of the Works but does not include materials ... intended to form ... the permanent work or vehicles engaged in transporting ..." The definition of "Construction Plant " is very wide and should include Temporary Works such as scaffolding or even likely includes temporary buildings though it is mentioned separately. The wording "owned by the Contractor" means that hired Plant, hire-purchase Plant or those Plants partly owned by the Contractor are excluded.

Clause 71 "... but may be removed from the Site by the Contractor at any time unless removal is expressly prohibited by the Engineer in writing. Upon removal ... shall re-vest in the Contractor ..."

This allows the Contractor to have a free hand to remove the plant unless prohibited by the Engineer in writing. As discussed in the commentary, the Contractor would always remove his expensive Plant when he anticipates critical situations, such as re-entry or bankruptcy and the Engineer may not be able to expressly prohibit his doing so in good time. The wording "... shall re-vest in the Contractor" suggests that the Plant, once removed, shall become the property of the Contractor and he is free to sell them.

Clause 72 "All materials owned by the Contractor for incorporation in the Works shall be and become the property of the Employer upon delivery to the Site, ..."

There is no difficulty to apply this Clause because the Contractor will be able to claim interim payment once the materials arrive on the Site. There may be areas which are not part of the Site where the Contractor uses for the manufacture of say, precast units or for storage of component parts, which would be brought to the Site for the Works. In Clause 1(1), the definition of the Site allows the Engineer to extend the Site to cover these areas. However, if such areas are outside Hong Kong, this may create complications because of the difference in applicable law or construction practice.

Comparison with the ICE Conditions

There is no provision in this ICE Clause 54 for Contractor's Equipment, Temporary Works materials or other goods or materials on the Site owned by the Contractor becoming the property of the Employer, except that those shall not be removed without the written consent of the Engineer whose consent shall not be unreasonably withheld. However, if the Contractor wishes to claim payment on goods or materials not yet arrived on the Site under Clause 60(1)(c), the Contractor may have to transfer to the Employer the property in such goods or materials. This will only happen when the Contract allows for it or agreed by the Employer and such goods or materials are substantially ready for incorporation into the Works. There is also the requirement to produce documentary evidence that the goods or materials have been vested in the Contractor by the supplier, suitably marked and set aside and stored to the satisfaction of the Engineer, all as detailed in sub-clauses (4) and (5).

Comparison with the FIDIC Conditions

This FIDIC Clause 4.17 is similar to the ICE Clause 54 in that the Contractor's Equipment when brought onto the Site shall be deemed to be exclusively intended for the execution of the Works. However, only major items of the Contractor's Equipment require the Engineer's consent for their removal.

There is an optional clause allowing the Contractor's Equipment owned by the Contractor to become the property of the Employer on its arrival on the Site. This presumably should be applicable to major Equipment.

 43 Can the Employer retain the falsework system and/or the crawler crane when the Contractor becomes bankrupt?

Clause 78

Contractor's Interim Statements

(1) The Contractor shall submit to the Engineer at the end of each monthly period (the first of such monthly periods to commence on the date for commencement of the Works) a statement showing:

    (a) the estimated contract value of the work done in accordance with the Contract up to the end of such monthly period with sums payable in respect of Nominated Sub-contractors listed separately;

    (b) a list of materials delivered to the Site for use in the permanent work and their estimated contract value;

    (c) all further estimated sums which the Contractor considers to be due to him under the Contract.

(2) The statement shall be prepared on a form supplied by and at the expense of the Contractor and the style and number of copies shall be as the Engineer shall determine. The Contractor shall complete the required number of copies of the statement and deliver them to the Engineer for checking and, if necessary, correction in accordance with Clause 79. One corrected copy shall be returned to the Contractor.

Corresponding clauses in the ICE Conditions:

Clause 60

(1) The Contractor shall submit to the Engineer at monthly intervals a statement (in such form if any as may be prescribed in the Specification) showing

(a) the estimated contract value of the Permanent Works executed up to the end of that month

(b) a list of any goods or materials delivered to the Site for but not yet incorporated in the Permanent Works and their value

(c) a list of any of those goods or materials identified in the Appendix to the Form of Tender which have not yet been delivered to the Site but of which the property has vested in the Employer pursuant to Clause 54 and their value and

(d) the estimated amounts to which the Contractor considers himself entitled in connection with all other matters for which provision is made under the Contract including any Temporary Works or Contractor's Equipment for which separate amounts are included in the Bill of Quantities

unless in the opinion of the Contractor such values and amounts together will not justify the issue of an interim certificate.

Amounts payable in respect of Nominated Sub-contracts are to be listed separately.

Corresponding clauses in the FIDIC Conditions:

Clause 14.3

The Contractor shall submit a Statement in six copies to the Engineer after the end of each month, in a form approved by the Engineer, showing in detail the amounts to which the Contractor considers himself to be entitled, together with supporting documents which shall include the report on the progress during this month in accordance with Sub-Clause 4.21 [*Progress Reports*].

The Statement shall include the following items, as applicable, which shall be expressed in the various currencies in which the Contract Price is payable, in the sequence listed:

(a) the estimated contract value of the Works executed and the Contractor's Documents produced up to the end of the month (including Variations but excluding items described in sub-paragraphs (b) to (g) below);

(b) any amounts to be added and deducted for changes in legislation and changes in cost, in accordance with Sub-Clause 13.7 [*Adjustments for Changes in Legislation*] and Sub-Clause 13.8 [*Adjustments for Changes in Cost*];

(c) any amount to be deducted for retention, calculated by applying the percentage of retention stated in the Appendix to Tender to the total of the above amounts, until the amount so retained by the Employer reaches the limit of Retention Money (if any) stated in the Appendix to Tender;

(d) any amounts to be added and deducted for the advance payment and repayments in accordance with Sub-Clause 14.2 [*Advance Payment*];

(e) any amounts to be added and deducted for Plant and Materials in accordance with Sub-Clause 14.5 [*Plant and Materials Intended for the Works*];

(f)　any other additions or deductions which may have become due under the Contract or otherwise, including those under Clause 20 [*Claims, Disputes and Arbitration*]; and

(g)　the deduction of amounts certified in all previous Payment Certificates.

條款 78　香港特別行政區土木工程合同一般條件：

「承建商」的中期報表

(1)　「承建商」必須於每個月結算期（首月結算期由本工程開始那日起計）完結時向「工程師」提交報表，示明：

(a)　直至每月算結期完結為止已按照「合同」完成的工程的估計合約價值及分別列出須支付各「指定次承判商」的款項；

(b)　已送往「工地」以供「永久工程」所用的物料的清單及此等物料的估計合約價值；

(c)　「承建商」認為按照「合同」應獲得支付的所有另外的估計款項。

(2)　報表須以「承建商」自費供應的表格編製，其格式及副本份數由「工程師」決定。「承建商」必須填妥所需的報表副本份數並交給「工程師」查閱及如有需要根據第79條改正報表。「工程師」必須將一份已改正的正本交回「承建商」。

Commentary

This Clause 78 is relatively straightforward and describes the procedures by which the Contractor submits his statement of the work done or any other things entitling him to claim interim payment from the Employer. Apart from the works completed, the Contractor is entitled to be paid also the preliminary items, materials on the Site, contractual claims and variations. In drawing up his statement, the Contractor must follow a certain form usually stipulated in the Specification or as determined by the Engineer.

Analysis and application

No detailed analysis is required for this Clause.

Comparison with the ICE Conditions

The ICE Clause 60(1)(c) allows the Contractor to claim payment for any goods or materials identified in the Appendix to the Form of Tender, which have not yet been delivered to the Site but may be stored or manufactured in other places, subject to the procedures and requirements given in Clause 54.

Comparison with the FIDIC Conditions

The FIDIC Clause 14.3 specifies the sequence of various items to be included in the Contractor's statement. The estimated contract value of the Works executed and the Contractor's Documents should be included first, followed by the amount of adjustment due to changes in legislation and changes in Cost, then the amount to be deducted for retention, the amount due to advance payment and repayment, Plant and materials intended for the Works, the claims and settlement sums and finally deduction of amounts certified in all previous Payment Certificates.

Clause 79

Interim and Final Payments, Retention Money and Interest

(1) Within 21 days (unless otherwise stated in the Contract) of the date of delivery to the Engineer of the Contractor's statement in accordance with Clause 78, the Engineer shall value and certify and within a further 21 days the Employer shall pay to the Contractor after deducting previous payments on account (if any) and any other sum deductible by the Employer under the Contract the sum which in the opinion of the Engineer is due, based on the rates in the Contract where appropriate, in respect of the following:

(a) the estimated value of the permanent work executed, and

(b) the estimated value of any Temporary Works or preliminary item for which a separate sum is provided in the Bills of Quantities, and

(c) the estimated value of materials for inclusion in the permanent work and not being prematurely delivered to and being properly stored on the Site, and

(d) the estimated sums payable in respect of Nominated Sub-contractors, and

(e) any other estimated sum to which, in the opinion of the Engineer, the Contractor is entitled in accordance with the Contract.

Provided that the total certified sum shall be adjusted by the Engineer to take into account:

(i) the retention of the percentage stated in the Contract until the sum retained reaches the limit of Retention Money stated in the Contract; and

(ii) any adjustment to be made for fluctuations in the cost of labour and materials in accordance with Clause 89.

Provided further that, for the purpose of interim payments, the value of the materials as referred to in (c) above for use in connection with any item of permanent work priced in the Contract shall be determined on the basis of the rate set out in the Contract for such work.

Corresponding clauses in the ICE Conditions:

Clause 60

(2) Within 25 days of the date of delivery of the Contractor's monthly statement to the Engineer or the Engineer's Representative in accordance with sub-clause (1) of this Clause the Engineer shall certify and within 28 days of the same date the Employer shall pay to the Contractor (after deducting any previous payments on account)

(a) the amount which in the opinion of the Engineer on the basis of the monthly statement is due to the Contractor on account of sub-clauses (1)(a) and (1)(d) of this Clause less a retention as provided in sub-clause (5) of this Clause and

(b) such amounts (if any) as the Engineer may consider proper (but in no case exceeding the percentage of the value stated in the Appendix to the Form of Tender) in respect of sub-clauses (1)(b) and (1)(c) of this Clause.

The payments become due on certification with the final date for payment being 28 days after the date of delivery of the Contractor's monthly statement.

The amounts certified in respect of Nominated Sub-contracts shall be shown separately in the certificate.

Corresponding clauses in the FIDIC Conditions:

Clause 14.7

The Employer shall pay to the Contractor:

(a) the first instalment of the advance payment within 42 days after issuing the Letter of Acceptance or within 21 days after receiving the documents in accordance with Sub-Clause 4.2 [*Performance Security*] and Sub-Clause 14.2 [*Advance Payment*], whichever is later;

(b) the amount certified in each Interim Payment Certificate within 56 days after the Engineer receives the Statement and supporting documents; and

(c) the amount certified in the Final Payment Certificate within 56 days after the Employer receives this Payment Certificate.

Payment of the amount due in each currency shall be made into the bank account, nominated by the Contractor, in the payment country (for this currency) specified in the Contract.

Clause 14.5

If this Sub-Clause applies, Interim Payment Certificates shall include, under sub-paragraph (e) of Sub-Clause 14.3, (i) an amount for Plant and Materials which have been sent to the Site for incorporation in the Permanent Works, and (ii) a reduction when the contract value of such Plant and Materials is included as part of the Permanent Works under sub-paragraph (a) of Sub-Clause 14.3 [*Application for Interim Payment Certificates*].

If the lists referred to in sub-paragraphs (b)(i) or (c)(i) below are not included in the Appendix to Tender, this Sub-Clause shall not apply.

The Engineer shall determine and certify each addition if the following conditions are satisfied:

(a) the Contractor has:

(i) kept satisfactory records (including the orders, receipts, Costs and use of Plant and Materials) which are available for inspection and

(ii) submitted a statement of the Cost of acquiring and delivering the Plant and Materials to the Site, supported by satisfactory evidence.

條款 79

香港特別行政區土木工程合同一般條件:

中期及末期付款、保留金及利息

(1) 「承建商」按照第78條交付報表給「工程師」的日期後的21天內（除非「合同」內另有說明），「工程師」必須對他認為到期須付的金額依據「合同」內所訂的價率(如適當的話)就下列各項釐訂價值及核實，而「僱主」必須在接着的21天內支付「承建商」該金額，但扣除之前已暫付的金額（如有的話）及「僱主」按照「合同」可扣除的任何其他金額。

(a) 已實施的「永久工程」的估計價值; 及

(b) 「工程量清單」另有列出款項的任何「臨時工程」或開辦項目的估計價值; 及

(c) 沒有過早送往「工地」及妥善地存放供「永久工程」使用的物料的估計價值; 及

(d) 須支付「指定次承判商」的估計金額; 及

(e) 「工程師」認為按照「合同」「承建商」有權獲得支付的任何其他估計金額。

但是，已核實的總金額必須由「工程師」對以下加以調整:

(i) 保留「合同」內說明的百分比的金額直至所保留的金額達到「合同」內說明的保留金限額為止; 及

(ii) 根據第89條規定任何因勞工及物料成本的浮動而作出的調整。

再者，為中期付款的目的而言，上述(c)段所提及與任何「合同」內已標價的「永久工程」項目有關而使用的物料的價值必須按照「合同」內所列出此等工程的價率釐訂。

Commentary

This Clause governs what kind of interim payment can be made to the Contractor upon his application, when and how it is made, how payments are subject to retention by the Employer and adjustment due to price fluctuations, when the retention money is released, what interest is to be paid in case of failure to make payment when it is due, the power of the Engineer to omit any value of work which he is not satisfied or delete, correct or modify any sum previously certified by him for proper reasons, and finally when and how to effect final payment. Some of the common situations which may be encountered are the claim for material on Site or off-Site,[1] on-account payments particularly for claims and variations, how adjustment items are to be made, the problem with high value preliminary items, payment for breach of contract and under certification or over payment.

Analysis and application

This sub-clause (1) above sets out the time limit of 21 days from the submission of statement by the Contractor within which the Engineer must value and then certify the entitlement of the Contractor under sub-heads (a) to (e). After adjustment due to retention and price fluctuation and deduction of previous payments, the Employer must then pay the amount due within another 21 days. Under sub-head (b), if a sum is allowed for Temporary Works or other preliminary items such as Contractor's office or third party insurance, it is necessary to assess what percentage of that sum is due. Usually the Contractor puts in high value for such items and the Engineer must then make an assessment, which is fair and acceptable to both parties. Under sub-head (c) regarding payment of material on the Site, it should be noted that the definitions of the Site is now extended to areas provided by the Contractor and agreed by the Engineer as forming part of the Site. Thus, the Contractor's own storage areas, say for precast units, can be forming part of the Site. The effect and consequence must be fully investigated if the Engineer agrees to extend the Site to outside the HKSAR territorial limit, especially concerning applicable law and construction practice, e.g., safety standards and environmental issues. Under sub-head (e), on-account payment for claims and variations can be made.

Interim and Final Payments, Retention Money and Interest

(2)   The Engineer may refuse to issue a certificate for an interim payment for a sum less than the minimum payment stated in the Contract, but nothing in this Clause shall prevent the Engineer from issuing a certificate at any time for any sum if in the opinion of the Engineer it is desirable to do so.

Corresponding clauses in the ICE Conditions:

Clause 60 (*continued*)

(3)   Until the whole of the Works has been certified as substantially complete in accordance with Clause 48 the Engineer shall not be bound to issue an interim certificate for a sum less than that stated in the Appendix to the Form of Tender but thereafter he shall be bound to do so and the certification and payment of amounts due to the Contractor shall be in accordance with the time limits contained in this Clause.

(4)   (*Quoted on p. 190.*)

(5)   The retention to be made pursuant to sub-clause (2) (a) of this Clause shall be the difference between

(a)   an amount calculated at the rate indicated in and up to the limit set out in the Appendix to the Form of Tender upon the amount due to the Contractor on account of sub-clauses (1)(a) and (1)(d) of this Clause and

(b)   any payment which shall have become due under sub-clause (6) of this Clause.

Corresponding clauses in the FIDIC Conditions:

Clause 14.5 (*continued*)

and either;

(b)   the relevant Plant and Materials:

(i)   are those listed in the Appendix to Tender for payment when shipped,

(ii)   have been shipped to the Country, en route to the Site, in accordance with the Contract; and

(iii)   are described in a clean shipped bill of lading or other evidence of shipment, which has been submitted to the Engineer together with evidence of payment of freight and insurance, any other documents reasonably required, and a bank guaranteed in a form and issued by an entity approved by the Employer in amounts and currencies equal to the amount due under this Sub-Clause: this guarantee may be in a similar form to the form referred to in Sub-Clause 14.2 [*Advance Payment*] and shall be valid until the Plant and Materials are properly stored on Site and protected against loss, damage or deterioration;

or

(c)   the relevant Plant and Materials:

(i)   are those listed in the Appendix to Tender for payment when delivered to the Site, and

(ii)   have been delivered to and are properly stored on the Site, are protected against loss, damage or deterioration, and appear to be in accordance with the Contract.

The additional amount to be certified shall be the equivalent of eighty percent of the Engineer's determination of the cost of the Plant and Materials (including delivery to the Site), taking account of the documents mentioned in this Sub-Clause and of the contract value of the Plant and Materials.

The currencies for this additional amount shall be the same as those in which payment will become due when the contract value is included under sub-paragraph (a) of Sub-Clause 14.3 [*Application for Interim Payment Certificates*]. At that time, the Payment Certificate shall include the applicable reduction which shall be equivalent to, and in the same currencies and proportions as, this additional amount for the relevant Plant and Materials.

Clause 14.6, paragraphs 1 and 2

No amount will be certified or paid until the Employer has received and approved the Performance Security. Thereafter, the Engineer shall, within 28 days after receiving a Statement and supporting documents, issue to the Employer an Interim Payment Certificate which shall state the amount which the Engineer fairly determines to be due, with supporting particulars.

However, prior to issuing the Taking-Over Certificate for the Works the Engineer shall not be bound to issue an Interim Payment Certificate in an amount which would (after retention and other deductions) be less than the minimum amount of Interim Payment Certificates (if any) stated in the Appendix to Tender. In this event, the Engineer shall give notice to the Contractor accordingly.

條款 79 (續)

香港特別行政區土木工程合同一般條件：

中期及末期付款、保留金及利息

(2) 「工程師」可以拒絕為少於在「合同」內說明的最低付款額的金額頒發中期付款證明書，如在「工程師」認為應當的情況下，本條不能阻止他隨時為任何金額頒發證明書。

Analysis and application

This sub-clause (2) may be applicable to the Engineer certifying a claim to be paid on a date not necessarily coinciding with the end of the monthly period or giving an additional certificate at the end of the financial year. These can be in addition to the monthly certificate.

 This steel deck unit is cast in the Mainland China. How can the Contractor claim interim payment? Would it be easier if the ICE Conditions were used?

Interim and Final Payments, Retention Money and Interest

(continued)

(3) Within 14 days of the date of issue by the Engineer of the maintenance certificate in accordance with Clause 80, the Engineer shall issue a certificate for the payment of Retention Money, which certificate shall state any Retention Money due to any Nominated Sub-contractor and, subject to Clause 83, the Employer shall pay such Retention Money to the Contractor within 21 days of the date of such certificate.

Corresponding clauses in the ICE Conditions:

Clause 60

(6) (a) Upon the issue of a Certificate of Substantial Completion in respect of any Section or part of the Works there shall become due to the Contractor one half of such proportion of the amount calculated to date under sub-clause (5)(a) of this Clause as the value of the Section or part bears to the value of the whole of the Works completed to date as certified under sub-clause (2)(a) of this Clause.

The total of the payments which shall become due under this sub-clause shall in no event exceed one half of the limit of retention set out in the Appendix to the Form of Tender.

(b) Upon issue of the Certificate of Substantial Completion in respect of the whole of the Works there shall become due to the Contractor one half of the amount calculated in accordance with sub-clause (5)(a) of this Clause less any payments which shall have become due under sub-clause (6)(a) of this Clause. Within 10 days of the date of issue of the said Certificate the Engineer shall certify the amount due.

Payment becomes due on certification of the amount due with the final date for payment being 14 days after the issue of the Certificate of Substantial Completion.

(c) At the end of the Defects Correction Period or if more than one the last of such periods there shall become due to the Contractor the remainder of the retention money.

Within 10 days of the date of the end of the said period the Engineer shall certify the amount due.

Payment becomes due on certification of the amount due with the final date for payment being 14 days after the end of the said period notwithstanding that at that time there may be outstanding claims by the Contractor against the Employer.

Provided that if at that time there remains to be carried out by the Contractor any outstanding work referred to under Clause 48 or any work ordered pursuant to Clauses 49 or 50 the Engineer may withhold certification until the completion of such work of so much of the said remainder as shall in the opinion of the Engineer represent the cost of the work remaining to be carried out.

Corresponding clauses in the FIDIC Conditions:

Clause 14.9

When the Taking-Over Certificate has been issued for the Works, the first half of the Retention Money shall be certified by the Engineer for payment to the Contractor. If a Taking-Over Certificate is issued for a Section or part of the Works, a proportion of the Retention Money shall be certified and paid. This proportion shall be two-fifths (40%) of the proportion calculated by dividing the estimated contract value of the Section or part, by the estimated final Contract Price.

Promptly after the latest of the expiry dates of the Defects Notification Periods, the outstanding balance of the Retention Money shall be certified by the Engineer for payment to the Contractor. If a Taking-Over Certificate was issued for a Section, a proportion of the second half of the Retention Money shall be certified and paid promptly after the expiry date of the Defects Notification Period for the Section. This proportion shall be two-fifths (40%) of the proportion calculated by dividing the estimated contract value of the Section by the estimate final Contract Price.

However, if any work remains to be executed under Clause 11 [*Defects Liability*], the Engineer shall be entitled to withhold certification of the estimated cost of this work until it has been executed.

When calculating these proportions, no account shall be taken of any adjustments under Sub-Clause 13.7 [*Adjustments for Changes in Legislation*] and Sub-Clause 13.8 [*Adjustments for Changes in Cost*].

中期及末期付款、保留金及利息

(3) 「工程師」根據第80條頒發保養證明書的日期後14天內必須頒發支付「保留金」的證明書，而這「保留金」證明書必須說明應支付任何「指定次承判商」的任何「保留金」及除第83條另有規定外，「僱主」必須在這「保留金」證明書的日期後21天內支付這「保留金」給「承建商」。

Analysis and application

This sub-clause (3) enables the Engineer to issue a certificate for the release of Retention Money 14 days after the issue of the maintenance certificate. It should be noted that under Clause 80, the maintenance certificate shall only be issued after the completion of all outstanding work and all necessary remedial or rectification work which may take a few additional months or even a couple of years after the Maintenance Period as described in the commentary to Clause 58. The certificate for release of Retention Money shall state any Retention Money due to Nominated Sub-contractors. Unless there is money to be recovered from the Contractor under Clause 83, the Retention Money shall be paid within 21 days.

 The Contractor has to invest a lot of money in this launching girder. Can he obtain any interim payment under the HKSAR Conditions? How about if the ICE Conditions or the FIDIC Conditions are used?

Interim and Final Payments, Retention Money and Interest

(4) (a) In the event of failure by the Employer to pay the Contractor in compliance with the provisions of this Clause, the Employer shall pay to the Contractor interest at one percent below the judgment debt rate prescribed from time to time by *The Rules of the High Court* (Cap. 4A) upon any overdue payment from but not including the date on which the same should have been made.

(b) The Employer shall not under any circumstances be liable to pay to the Contractor interest on any sum payable to the Contractor under or arising out of the Contract, whether upon the certificate of the Engineer or otherwise, at a rate in excess of one percent below the said judgment debt rate.

Corresponding clauses in the ICE Conditions:

Clause 60

(7) In the event of

(a) failure by the Engineer to certify or the Employer to make payment in accordance with sub-clauses (2) (4) or (6) of this Clause or

(b) any decision of an adjudicator or any finding of an arbitrator to such effect

the Employer shall pay to the Contractor interest compounded monthly for each day on which any payment is overdue or which should have been certified and paid at a rate equivalent to 2% per annum above the base lending rate of the bank specified in the Appendix to the Form of Tender.

If in an arbitration pursuant to Clause 66 the arbitrator holds that any sum or additional sum should have been certified by a particular date in accordance with the aforementioned sub-clauses but was not so certified this shall be regarded for the purposes of this sub-clauses as a failure to certify such sum or additional sum. Such sum or additional sum shall be regarded as overdue for payment 28 days after the date by which the arbitrator holds that the Engineer should have certified the sum or if no such date is identified by the arbitrator shall be regarded as overdue for payment from the date of the Certificate of Substantial Completion for the whole of the Works.

Corresponding clauses in the FIDIC Conditions:

Clause 14.8

If the Contractor does not receive payment in accordance with Sub-Clause 14.7 [*Payment*], the Contractor shall be entitled to receive financing charges compounded monthly on the amount unpaid during the period of delay. This period shall be deemed to commence on the date for payment specified in Sub-Clause 14.7 [*Payment*], irrespective (in the case of its sub-paragraph (b)) of the date on which any Interim Payment Certificate is issued.

Unless otherwise stated in the Particular Conditions, these financing charges shall be calculated at the annual rate of three percentage points above the discount rate of the central banking in the country of the currency of payment, and shall be paid in such currency.

The Contractor shall be entitled to this payment without formal notice or certification, and without prejudice to any other right or remedy.

條款 79	

中期及末期付款、保留金及利息

（續）

(4) (a) 如「僱主」未能遵從本條條文付款給「承建商」，「僱主」必須為逾期未付的款額向「承建商」支付利息，而利息由須付款的日期起計但不計當日，按《高等法院規則》（第4A章）為判定債項而不時訂定的利率減1厘計算。

(b) 在任何情況下「僱主」皆無法律責任按高於判定債項利率減一厘的利率支付「承建商」依據「合同」或因「合同」所引致須支付「承建商」的任何金額的利息，無論「工程師」有否為此金額頒發證明書。

Analysis and application

Sub-clause 4(a) applies when the Employer fails to pay the Contractor the amount due on time, which is 21 days after the Engineer's certification. If the Engineer under-certifies the amount due or there is a delay in certifying within 21 days of receipt of the Contractor's statement, this Clause may not apply directly but the Contractor may submit a claim which may still be considered by the Engineer along the principle in this sub-clause. The judgment debt rate can be found on the internet and is usually higher than the prime lending rate. Though not explicitly mentioned, simple interest is payable as in the case of failure to pay a judgment debt. Sub-clause 4(b) limits the rate of interest to one percent below the judgment debt rate, even if the Contractor can show that his rate of borrowing to finance his undertaking exceeds the former rate. As regards whether the Engineer may be in breach of his duty of care to the Contractor who suffers loss as a result of the Engineer's failure to certify correctly and timely, the Court held that the Engineer would not normally be liable.[2]

Interim and Final Payments, Retention Money and Interest

(5) The Engineer shall have the power to omit from any certificate the value of any work done, materials supplied or services rendered with which the Engineer may for the time being be dissatisfied and for that purpose, or for any other reason which to the Engineer may seem proper, may by any certificate delete, correct or modify any sum previously certified by him.

Corresponding clauses in the ICE Conditions:

Clause 60

(8) The Engineer shall have power to omit from any certificate the value of any work done goods or materials supplied or services rendered with which he may for the time being be dissatisfied and for that purpose or for any other reason which to him may seem proper may by any certificate delete correct or modify any sum previously certified by him. Provided that

(a) the Engineer shall not in any interim certificate delete or reduce any sum previously certified in respect of work done goods or materials supplied or services rendered by a Nominated Sub-contractor if the Contractor shall have already paid or be bound to pay that sum to the Nominated Sub-contractor and

(b) if the Engineer in the final certificate shall delete or reduce any sum previously certified in respect of work done goods or materials supplied or services rendered by a Nominated Sub-contractor which sum shall have been already paid by the Contractor to the Nominated Sub-contractor the Employer shall reimburse to the Contractor the amount of any sum overpaid by the Contractor to the Sub-contractor in accordance with the certificate issued under sub-clause (2) of this Clause which the Contractor shall be unable to recover from the Nominated Sub-contractor together with interest thereon at the rate stated in sub-clause (7) of this Clause from 28 days after the date of the final certificate issued under sub-clause (4) of this Clause until the date of such reimbursement.

(9) Every certificate issued by the Engineer pursuant to this Clause shall be sent to the Employer and on the Employer's behalf to the Contractor. By this certificate the Employer shall give notice to the Contractor specifying the amount (if any) of the payment proposed to be made and the basis on which it was calculated.

(10) Where a payment under Clause 60(2) (4) or (6) is to differ from that certified or the Employer is to withhold payment after the final date for payment of a sum due under the Contract the Employer shall notify the Contractor in writing not less than one day before the final date for payment specifying the amount proposed to be withheld and the ground for withholding payment or if there is more than one ground each ground and the amount attributable to it.

Corresponding clauses in the FIDIC Conditions:

Clause 14.6

14.6.3 An Interim Payment Certificate shall not be withheld for any other reason, although:

(a) if any thing supplied or work done by the Contractor is not in accordance with the Contract, the cost of rectification or replacement my be withheld until rectification or replacement has been completed; and/or

(b) if the Contractor was or is failing to perform any work or obligation in accordance with the Contract, and had been so notified by the Engineer, the value of this work or obligation may be withheld until the work or obligation has been performed.

14.6.4 The Engineer may in any Payment Certificate make any correction or modification that should properly be made to any previous Payment Certificate. A Payment Certificate shall not be deemed to indicate the Engineer's acceptance, approval, consent or satisfaction.

條款 **79** (續)

中期及末期付款、保留金及利息

(5) 如「工程師」在某時候不滿意任何已完成的工程、供應的物料或提供的服務，他有權將這些工程、物料或服務的價值不包括在證明書內，及為此或為他認為適當的任何其他原因，「工程師」可憑着任何證明書刪除、更正或修改任何以前經他核實的金額。

Analysis and application

The first part of this sub-clause (5) enables the Engineer to withhold from any certificate the value of work carried out if the work is found to be unsatisfactory. The same is applicable to materials supplied and services rendered. The second part of this sub-clause empowers the Engineer to correct any previous certificates for proper reasons, such as over-certification, or materials or works subsequently found to be unsatisfactory.

 The completion of this viaduct has now overrun by two months. The Employer does not want to deduct liquidated damages during this critical stage of the Works. What advice can the Engineer give to the Employer?

(6) Within 90 days after the date of issue of the maintenance certificate the Contractor shall submit to the Engineer a statement of final account and supporting documentation showing in detail the value in accordance with the Contract of the work done in accordance with the Contract together with all further sums which the Contractor considers to be due to him under the Contract up to the date of the maintenance certificate. Within 90 days after receipt of the final account and of all information reasonably required for its verification, the Engineer shall issue a final payment certificate stating the sum which in his opinion is finally due under the Contract up to the date of the maintenance certificate, and after giving credit to the Employer for all sums previously paid by the Employer and for all sums to which the Employer is entitled under the Contract the Engineer shall state the balance (if any) due from the Employer to the Contractor or from the Contractor to the Employer as the case may be. Such balance shall be paid to or by the Contractor as the case may require within 28 days of the date of the certificate.

Corresponding clauses in the ICE Conditions:

Clause 60

(4) Not later than 3 months after the date of the Defects Correction Certificate the Contractor shall submit to the Engineer a statement of final account and supporting documentation showing in detail the value in accordance with the Contract of the Works carried out together with all further sums which the Contractor considers to be due to him under the Contract up to the date of the Defects Correction Certificate.

Within 3 months after receipt of this final account and of all information reasonably required for its verification the Engineer shall issue a certificate stating the amount which in his opinion is finally due under the Contract from the Employer to the Contractor or from the Contractor to the Employer as the case may be up to the date of the Defects Correction Certificate and after giving credit to the Employer for all amounts previously paid by the Employer and for all sums to which the Employer is entitled under the Contract.

Such amount shall subject to Clause 47 be paid to or by the Contractor as the case may require. The payment becomes due on certification. The final date for payment is 28 days later.

Corresponding clauses in the FIDIC Conditions:

Clause 14.10

Within 84 days after receiving the Taking-Over Certificate for the Works, the Contractor shall submit to the Engineer six copies of a Statement at completion with supporting documents, in accordance with Sub-Clause 14.3 [*Application for Interim Payment Certificates*], showing:

(a) the value of all work done in accordance with the Contract up to the date stated in the Taking-Over Certificate for the Works,

(b) any further sums which the Contractor considers to be due, and

(c) an estimate of any other amounts which the Contractor considers will become due to him under the Contract. Estimated amounts shall be shown separately in this Statement at completion.

The Engineer shall then certify in accordance with Sub-Clause 14.6 [*Issue of Interim Payment Certificates*].

Clause 14.11

Within 56 days after receiving the Performance Certificate, the Contractor shall submit, to the Engineer, six copies of a draft final statement with supporting documents showing in detail in a form approved by the Engineer:

(a) the value of all work done in accordance with the Contract, and

(b) any further sums which the Contractor considers to be due to him under the Contract or otherwise.

If the Engineer disagrees with or cannot verify any part of the draft final statement, the Contractor shall submit such further information as the Engineer may reasonably require and shall make such changes in the draft as may be agreed between them. The Contractor shall then prepare and submit to the Engineer the final statement as agreed. This agreed statement is referred to in these Conditions as the "Final Statement."

However if, following discussions between the Engineer and the Contractor and any changes to the draft final statement which are agreed, it becomes evident that a dispute exists, the Engineer shall deliver to the Employer (with a copy to the Contractor) an Interim Payment Certificate for the agreed parts of the draft final statement. Thereafter, if the dispute is finally resolved under Sub-Clause 20.4 [*Obtaining Dispute Adjudication Board's Decision*] or Sub-Clause 20.5 [*Amicable Settlement*], the Contractor shall then prepare and submit to the Employer (with a copy to the Engineer) a Final Statement.

條款 79 (續)

香港特別行政區土木工程合同一般條件：

中期及末期付款、保留金及利息

(6) 「承建商」必須在保養證明書頒發的日期後90天內提交最終帳目報表及支持文件給「工程師」，詳細標明「承建商」按照「合同」完成的工程而依據「合同」計算的工程價值，及「承建商」認為直至保養證明書的日期為止按照「合同」應獲得支付的所有其他金額。「工程師」在收到最終帳目及為核實這帳目而合理地需要的所有資料的90天內必須頒發最終付款證明書，說明他認為直至保養證明書的日期為止按照「合同」最終須付的金額。「工程師」在扣除「僱主」之前已支付的所有金額及按照「合同」「僱主」有權獲得支付的所有金額後，必須說明「僱主」須支付「承建商」或「承建商」須支付「僱主」（視情況而定）的餘額（如有的話）。此餘額必須在該最終付款證明書的日期後28天內由「僱主」支付「承建商」或由「承建商」支付「僱主」（視情況而定）。

Analysis and application

This sub-clause (6) describes the procedures and the obligation of the Contractor and the Engineer in finalizing the account after the issue of the maintenance certificate. As explained before, the maintenance certificate is usually issued many months or even a couple of years after the expiry of the Maintenance Period and by that time, all the valuation of variations and most of the claims should have been finalized and the retention money has been released. There should be very little money due to the Contractor or vice versa. However, there may still be one or a few outstanding claims, which may have to be decided by the Engineer or to be resolved by mediation or arbitration. In practice, the Contractor may not observe this 90 days time limit either because not all documents are readily available or there is a lack of manpower to gather and submit information. There is little sanction against the Contractor except that the Engineer may unilaterally issue the final payment certificate under sub-clause (7). However, the Engineer must observe his 90 days time limit to issue the final payment certificate or the Contractor may be able to claim interest. Also, after the issue of the final payment certificate, the Employer must pay within 28 days or Clause 79(4)(a) regarding interest on late payment shall apply.

The HKSAR Conditions:

Interim and Final Payments, Retention Money and Interest

(7) If the Contractor fails to submit a statement of final account within 90 days of the date of the maintenance certificate in accordance with sub-clause (6) of this Clause the Engineer shall be entitled to issue a final payment certificate without reference to the Contractor.

Corresponding clauses in the ICE Conditions:

(There is no corresponding sub-clause.)

Corresponding clauses in the FIDIC Conditions:

Clause 14.12

When submitting the Final Statement, the Contractor shall submit a written discharge which confirms that the total of the Final Statement represents full and final settlement of all moneys due to the Contractor under or in connection with the Contract. This discharge may state that it becomes effective when the Contractor has received the Performance Security and the outstanding balance of this total, in which event the discharge shall be effective on such date.

Clause 14.13

Within 28 days after receiving the Final Statement and written discharge in accordance with Sub-Clause 14.11 *[Application for Final Payment Certificate]* and Sub-Clause 14.12 *[Discharge]*, the Engineer shall issue, to the Employer, the Final Payment Certificate which shall state:

(a) the amount which is finally due, and

(b) after giving credit to the Employer for all amounts previously paid by the Employer and for all sums to which the Employer is entitled, the balance (if any) due from the Employer to the Contractor or from the Contractor to the Employer, as the case may be.

If the Contractor has not applied for a Final Payment Certificate in accordance with Sub-Clause 14.11 *[Application for Final Payment Certificate]* and Sub-Clause 14.12 *[Discharge]*, the Engineer shall request the Contractor to do so. If the Contractor fails to submit an application within a period of 28 days, the Engineer shall issue the Final Payment Certificate for such amount as he fairly determines to be due.

Notes

1. For a general discussion on payment of materials off-Site, see Roger Knowles, *150 Contractual Problems and Their Solutions*, Note 7.3 .

2. *Pacific Associates v Baxter* [1990] 1 QB993 CA; *Leon Engineering and Construction Co. Ltd. v Ka Duk Investment Co.* [1989] 47 BLR139.

中期及末期付款、保留金及利息

（續）

(7) 如「承建商」未能按照本條第(6)款在保養證明書的日期90天內提交最終帳目報表，「工程師」有權在無須諮詢「承判商」下頒發最終付款證明書。

Analysis and application

In practice, the Engineer will allow adequate time for the Contractor to arrange the submission of his final account. The issue of a final payment certificate without consultation with the Contractor under this sub-clause (7) is undesirable because such a final payment certificate may still be disputed by the Contractor.

Comparison with the ICE Conditions

The main differences between this ICE Clause 60 and the HKSAR Clause 79 are:

(i) The ICE sub-clause 60(2) requires payment from the Employer only 28 days after the delivery of monthly statement from the Contractor while the HKSAR Clause allows 21 days plus 21 days, i.e., a total of 42 days. The ICE sub-clause allows 25 days for the Engineer to certify but only 3 more days for the Employer to make payment.

(ii) The goods or materials under the ICE Clause 60(1)(b) and (c) are not subject to retention. Clause 60(1)(c) allows off-Site payment of materials[1] specified in Appendix to Form of Tender.

(iii) Under the ICE Clause, one-half of the retention money is released to the Contractor upon the issue of the Certificate of Substantial Completion of the Works, or one-half of the proportional amount of retention money is released to the Contractor upon the issue of the Certificate of Substantial Completion of any Section or part of the Works, depending on whether the Works are divided into Sections or not. The remaining retention money is released to the Contractor at the end of the Defects Correction Period or the last of the Defects Correction Periods. Under the HKSAR Clause, retention money will only be released after the issue of the maintenance certificate.

(iv) Under the ICE sub-clause 60(7), the interest of any overdue payment is compounded monthly at a rate of 2% above the best lending rate, while under the HKSAR Clause, simple interest is usually payable at 1% below the judgment debt rate.

(v) Under the ICE sub-clause 60(8), it is explicitly stated that the power of the Engineer to omit from any certificate the value of materials or works which are found to be unsatisfactory shall not extend to works or materials of the Nominated Sub-contractors.

Comparison with the FIDIC Conditions

The main differences between the FIDIC Clause 14.7 and the HKSAR Clause 79 are:

(i) The FIDIC Clause 14.7 allows for advance payment to the Contractor as an interest-free loan for mobilization.

(ii) The total time allowed for the Employer to make payment certified by the Engineer is 56 days after receiving the Contractor's statement.

(iii) The FIDIC sub-clause 14.5(b) allows for Plant and Material listed in Appendix to Tender to be paid once shipped or have been shipped to the Country en route to the Site.

(iv) Similar to the ICE Clause, the FIDIC Clause 14.9 allows the release of one-half of the retention money or proportional money after the issue of the Taking-Over Certificate. It also allows further release of remaining retention money after the expiry of the Defects Notification Period.

(v) The FIDIC Clause 14.8 also allows finance charge to be compounded monthly for the Employer's failure to make payment.

(vi) The FIDIC Clause 14.6 takes a more cautious approach for withholding the Interim Payment Certificate on unsatisfactory works.

(vii) The FIDIC Clause gives a more detailed procedure to follow for the Final Payment Certificate and especially in case of disagreement by the Contractor or a dispute arises between the Employer/Engineer and the Contractor.

(Please see opposite page for Notes.)

Clause 80

Maintenance Certificate

(1) Upon the expiry of the Maintenance Period, or where there is more than one such Period, upon the expiry of the latest Period and when all outstanding work referred to under Clause 53 and all work of repair, reconstruction, rectification and making good any defect, imperfection, shrinkage and other fault referred to in Clause 56 shall have been completed the Engineer shall issue a maintenance certificate stating the date on which the Contractor shall have completed his obligation to execute the Works.

(2) No certificate, other than the maintenance certificate, shall be deemed to constitute approval of any work or other matter in respect of which it is issued or shall be taken as an admission of the due performance of the Contract or any part thereof.

   Provided that the maintenance certificate shall not be deemed to constitute approval of any work or other matter in respect of which it is issued which has not been carried out in accordance with the Contract and which the Engineer could not with reasonable diligence have discovered before the issue of the maintenance certificate.

(3) The issue of any certificate including the maintenance certificate shall not be taken as relieving either the Contractor or the Employer from any liability the one towards the other arising out of or in any way connected with the performance of their respective obligations under the Contract. Provided that the Employer shall not be liable to the Contractor for any matter or thing arising out of or in connection with the Contract or the execution of the Works unless the Contractor shall have made a claim in relation thereto in accordance with the time limits specified in Clause 50 or Clause 64.

Corresponding clauses in the ICE Conditions:

Clause 61

(1) At the end of the Defects Correction Period or if more than one the last of such periods and when all outstanding work referred to under Clause 48 and all work of repair amendment reconstruction rectification and making good of defects imperfections shrinkages and other faults referred to under Clauses 49 and 50 have been completed the Engineer shall issue to the Employer (with a copy to the Contractor) a Defects Correction Certificate stating the date on which the Contractor shall have completed his obligations to construct and complete the Works to the Engineer's satisfaction.

(2) The issue of the Defects Correction Certificate shall not be taken as relieving either the Contractor or the Employer from any liability the one towards the other arising out of or in any way connected with the performance of their respective obligations under the Contract.

Corresponding clauses in the FIDIC Conditions:

Clause 11.9

Performance of the Contractor's obligations shall not be considered to have been completed until the Engineer has issued the Performance Certificate to the Contractor, stating the date on which the Contractor completed his obligations under the Contract.

   The Engineer shall issue the Performance Certificate within 28 days after the latest of the expiry dates of the Defects Notification Periods, or as soon thereafter as the Contractor has supplied all the Contractor's Documents and completed and tested all the Works, including remedying any defects. A copy of the Performance Certificate shall be issued to the Employer.

   Only the Performance Certificate shall be deemed to constitute acceptance of the Works.

條款 80

保養證明書

(1) 在「保養期」（如有多於一個「保養期」，則指最後的一個）完結時及按照第53條所提及的所有仍未完成的工程及按照第56條所提及的所有要修理、重建、糾正、或修復的任何缺失、欠完善之處、縮減及其他錯誤的工程都完成後，「工程師」必須頒發保養證明書列明「承建商」已完成實施「本工程」的責任的日期。

(2) 除保養證明書外，其他為任何工程或事項所頒發的證明書皆不可被視為構成對此等工程或事項的批核，亦不可被視為對「合同」或其任何部分的妥善履行的承認。

但如任何工程或其他事項並未按照「合同」進行而「工程師」在頒發保養證明書之前無法在合理及盡力的情況下發現此事，則此保養證明書不可被視為構成對此等工程或事項的批核。

(3) 任何證明書，包括保養證明書的頒發皆不可被視為減免「承建商」或「僱主」在各自履行「合同」的責任上所引致的，或任何與此有關的，對另一方須負的任何法律責任。但除非「承建商」已按照第50或64條指明的時限提出索償，「僱主」無須對「承建商」就「合同」或「本工程」的實施所引致的或與此有關的任何事情或事物負上法律責任。

Commentary

The issue of maintenance certificate is important because it is linked to the release of retention money and the finalization of account. The Period of Maintenance is specified in the Appendix to the Form of Tender and usually lasts for 365 days for civil engineering contracts. As explained in the commentary to Clause 79, unlike the ICE Conditions and the FIDIC Conditions, the retention money is not released just upon the expiry of the Maintenance Period but only after the issue of the maintenance certificate upon completion of all outstanding works and works of repair, reconstruction, rectification and making good defects. For major civil engineering contracts, there are usually very substantial outstanding works to be completed as explained in the commentary to Clause 53. For civil engineering contracts with landscape works, the Establishment Period is usually the last in time and a portion of retention money is retained until its expiry. For civil engineering contracts with substantial E & M works, Special Conditions are required because the maintenance and rectification of defects in E & M components are of a different nature and the Maintenance Period can be extended and therefore the procedures for the release of retention money must be carefully considered. Apart from the retention money, the maintenance certificate has once raised an issue as to whether the certificate is conclusive and binding on the Employer so as to prevent him from further questioning of workmanship and materials.[1] The matter has now been fully clarified in sub-clause (2) and (3). There has also been an issue on whether the Engineer's power ends with the issue of the maintenance certificate and it has now been clarified that it does not[2] and in any case, the Engineer would still be involved in the preparation of the final account after the issue of the maintenance certificate. Finally, the issue of the maintenance certificate prevents the Contractor from raising any new claims as stated in the second part of sub-clause (3).

Analysis and application

Clause 80(1) "Upon the expiry of Maintenance Period ... and all work of repair, reconstruction, rectification and making good any defect, imperfection, shrinkage and other fault referred to in Clause 56 shall have been completed ..."

As described in the commentary, apart from the completion of outstanding works, the requirement to rectify all defects before the issue of the maintenance certificate (which is a precondition to the release of retention money) will hold up the Contractor's money for a long period. It should be noted that sometimes the maintenance authorities put very stringent requirements or additional requirements before accepting their respective parts of the Works for maintenance and it is submitted that such practice may attract claims from the Contractor and the Engineer must take a balanced approach on the authorities' requirements in handing over the Works.

Clause 80(1) "... the Engineer shall issue a maintenance certificate stating the date on which the Contractor shall have completed his obligation ..."

Though not expressly stated as in the case of the ICE Conditions and the FIDIC Conditions, the maintenance certificate is to be issued to the Employer and copied to the Contractor.

Clause 80(2) "No certificate, other than the maintenance certificate, shall be deemed to constitute approval of any works ... or shall be taken as an admission of the due performance of the Contractor ..."

In most cases, the maintenance certificate is the only certificate to conclude that the Contractor has duly performed his duties regarding workmanship and materials. Hence, the certificate of completion would not have such an effect. The conclusiveness of the maintenance certificate is, however, qualified by the proviso of sub-clause (2).

Clause 80(2) "... Provided that the maintenance certificate shall not be deemed to constitute approval of any work ... which the Engineer could not with reasonable diligence have discovered before the issue of the maintenance certificate."

The proviso is self-explanatory and the maintenance certificate will not cover any latent defects, which the Engineer could not with reasonable diligence have discovered. What constitutes reasonable diligence is a matter for the arbitrator or the Court to decide. It is advisable that the Engineer shall carry out a full and thorough inspection, preferably after a brief review of the Contract requirements, in order to satisfy the requirements of "reasonable diligence."

Clause 80(3) "The issue of any certificate including the maintenance certificate shall not be taken as relieving either the Contractor or the Employer from any liability the one towards the other ... with the performance of their respective obligations under the Contract ..."

From the Contractor's viewpoint, there are still claims to be pursued after the issue of the maintenance certificate. From the Employer's viewpoint, any latent defects discovered after the issue of the maintenance certificate would still be the liability of the Contractor.

Clause 80(3) "... Provided that the Employer shall not be liable ... unless the Contractor shall have made a claim in relation thereto in accordance with the time limits specified in Clause 50 or Clause 64."

Apparently, this proviso does not say anything apart from restating the principles in Clause 50 and 64. The practical effect may be that even though the Contractor has given due notice and submitted claim details as required under these clauses, the Contractor may not be able to submit any new facts to substantiate or further his claims after the issue of the maintenance certificate.

Comparison with the ICE Conditions

Apart from the term "Defects Correction Period" is used, the ICE sub-clause 61(1) is very similar to the HKSAR sub-clause 80(1) regarding the preconditions to the issue of the maintenance certificate, i.e., all outstanding works and rectification works are completed. The ICE Sub-clause 61(2) describes generally that the liability towards each other is not relieved upon the issue of the said certificate while the HKSAR sub-clause 80(2) mentions more specifically that materials and workmanship are not deemed to be approved upon issue of the maintenance certificate unless these could be discovered with reasonable diligence. The last sentence of the HKSAR sub-clause 80(3) appears to preclude the Contractor from submitting new claims or new facts about existing claim after the maintenance certificate is issued. It should also be noted that under the ICE Conditions, the release of retention money is not directly linked to the issue of the Defects Correction Certificate, unlike the HKSAR Conditions.

Comparison with the FIDIC Conditions

The FIDIC Clause 11.9 is similar to the HKSAR Clause and the ICE Clause in general. Though it states that "Only the Performance Certificate shall be deemed to constitute acceptance of the Works," the Contractor is still responsible for any latent defects.[3] Like the ICE Conditions, the release of retention money is not directly linked to the issue of the Performance Certificate.

Notes

1. Ian Duncan Wallace, *A Commentary on the FIDIC International Standard Form of Engineering and Building Contract*, p. 149.
2. Max Abrahamson, *Engineering Law and the ICE Contracts*, 4th Edition, p. 275.
3. See the commentary on Clauses 11.9 and 11.10 in Peter L. Booen (Principal drafter), *The FIDIC Contracts Guide*.

 Q 47 The Maintenance Period is over, and the Employer now discovers a lot of defects in the cladding of the building. Who should be responsible for the repair? What is the position after the maintenance certificate has been issued? Would it be different under the ICE Conditions or the FIDIC Conditions?

The HKSAR Conditions:

Determination of the Contractor's Employment

(1) If the Contractor shall become bankrupt or have a receiving order made against him or shall present his petition in bankruptcy or shall make an arrangement with or assignment in favour of his creditors or shall agree to carry out the Contract under a committee of inspection of his creditors or (being a corporation) shall go into liquidation (other than a voluntary liquidation for the purposes of amalgamation or reconstruction) or if the Contractor shall assign the Contract without the consent in writing of the Employer first obtained or shall have an execution levied on his goods or if the Engineer shall certify in writing to the Employer that in his opinion the Contractor:

   (a) has abandoned the Contract, or

   (b) without reasonable excuse has failed to commence the Works in accordance with Clause 47, or

   (c) has suspended the progress of the Works for 14 days after receiving from the Engineer notice in writing to proceed, or

   (d) has failed to comply with an order from the Engineer given in accordance with Clause 46, or

   (e) despite previous warning by the Engineer in writing is failing to proceed with the Works with due diligence or is persistently in breach of any of his obligations under the Contract, or

   (f) has sub-contracted the Works, or

   (g) has to the detriment of good workmanship or in defiance of the Engineer's instruction to the contrary sub-contracted any part of the Works, *(to be continued on page 202)*

Corresponding clauses in the ICE Conditions:

Clause 65

(1) In the event that the Contractor

   (a) assigns or attempts to assign the Contract or any part thereof or any benefit or interest thereunder without the prior written consent of the Employer or

   (b) is in breach of Clause 4(1) or

   (c) (i) becomes bankrupt or presents his petition in bankruptcy or

   (ii) has a receiving order or administration order made against him or

   (iii) makes an arrangement with or an assignment in favour of his creditors or

   (iv) agrees to carry out the Contract under a committee of inspection of his creditors or

   (v) (being a corporation) has a receiver or administrator appointed or goes into liquidation (other than a voluntary liquidation for the purposes of amalgamation or reconstruction) or

   (d) has an execution levied on his goods which is not stayed or discharged within 28 days

or if the Engineer certifies in writing to the Employer with a copy to the Contractor that in his opinion the Contractor

   (e) has abandoned the Contract without due cause or

   (f) without reasonable excuse has failed to commence the Works in accordance with Clause 41 or

   (g) has suspended the progress of the Works without due cause for 14 days after receiving from the Engineer written notice to proceed or

   (h) has failed to remove goods or materials from the Site or to pull down and replace work for 14 days after receiving from the Engineer written notice that the said goods materials or work has been condemned and rejected by the Engineer or

   (j) despite previous warnings by the Engineer in writing is failing to proceed with the Works with due diligence or is otherwise persistently or fundamentally in breach of his obligations under the Contract *(to be continued on page 202)*

Corresponding clauses in the FIDIC Conditions:

Clause 15.2

The Employer shall be entitled to terminate the Contract if the Contractor:

   (a) fails to comply with Sub-Clause 4.2 [*Performance Security*] or with a notice under Sub-Clause 15.1 [*Notice to Correct*],

(b) abandons the Works or otherwise plainly demonstrates the intention not to continue performance of his obligations under the Contract,

(c) without reasonable excuse fails:

(i) to proceed with the Works in accordance with Clause 8 [*Commencement, Delays and Suspension*], or

(ii) to comply with a notice issued under Sub-Clause 7.5 [*Rejection*] or Sub-Clause 7.6 [*Remedial Work*], within 28 days after receiving it,

(d) subcontracts the whole of the Works or assigns the Contract without the required agreement,

(e) becomes bankrupt or insolvent, goes into liquidation, has a receiving or administration order made against him, compounds with his creditors, or carries on business under a receiver, trustee or manager for the benefit of his creditors, or if any act is done or event occurs which

(under applicable Laws) has a similar effect to any of these acts or events, or

(f) gives or offers to give (directly or indirectly) to any person any bribe, gift, gratuity, commission or other thing of value, as an inducement or reward:

(i) for doing or forbearing to do any action in relation to the Contract, or

(ii) for showing or forbearing to show favour or disfavour to any person in relation to the Contract, or if any of the Contractor's Personnel, agents or Subcontractors gives or offers to give (directly or indirectly) to any person any such inducement or reward as is described in this sub-paragraph (f). However, lawful inducements and rewards to Contractor's Personnel shall not entitle termination.

(to be continued on page 203)

條款 81

香港特別行政區土木工程合同一般條件：

終止僱用「承建商」

(1) 如「承建商」將要破產或有接管令發給「承建商」或「承建商」將會提交破產申請或將會與其債權人作出安排或向其債權人作出轉讓債權或將會同意在其債權人的監督委員會之下履行「合同」或（如「承建商」是一法團）將會進行清盤（為合併或重組目的進行的自動清盤除外）或未經「僱主」書面同意而轉讓「合同」或其物品將被扣押或如「工程師」以書面向「僱主」核實他認為「承建商」：

(a) 已放棄「合同」；或

(b) 在沒有合理解釋下不按照第47條開始「本工程」；或

(c) 在已接獲「工程師」有關進行「本工程」的書面通知後擱置進行「本工程」長達14天；或

(d) 未能遵從「工程師」根據第46條發出的命令；或

(c) 雖然「工程師」之前已發出書面警告，但「承建商」仍未能盡力進行「本工程」或不斷地違反他在「合同」下的任何責任；或

(f) 已將「本工程」全部分包；或

(g) 在有損良好技術水平或違抗「工程師」指令不可將「本工程」某部分分包的情況下將「本工程」此部分分包，(頁203續)

Commentary

This Clause sets out the various prerequisites of the contractual right of determination by the Employer by way of entering upon the Site and expelling the Contractor. This contractual right is in addition to the Employer's right under the general law by which he can notify the Contractor that he is treating the Contract as an end when the Contractor is in serious or fundamental breach of the Contract. Under this Clause, the Employer only tells the Contractor to leave the Site and does not purport to end the Contract itself.[1] The Employer can continue with the Work either by his own workers or by using a new contractor but the

Engineer can control the state of accounts between the Employer and the defaulting Contractor by issuing a further certificate once the cost of completion is known. Under normal circumstances, liquidated damages can continue to be applied to cover the full period from the contractual completion date to the date of actual completion, even though for part of the time the work is out of the defaulting Contractor's control. However, the defaulting Contractor may be entitled to an extension of time for any unnecessary delay in getting a new contractor to take over the work or any delay by the new contractor.[2]

In exercising the right under this Clause, legal advice must be sought since any attempt to expel the Contractor from the Site not strictly in accordance with the provision of this Clause will amount to a repudiation of the Contract by the Employer, entitling the Contractor to give up the Contract and claim damages from the Employer. Other possible steps such as a supplemental agreement, novation, having part of the Works carried out by the Term Contractor, etc. could be considered, usually with the approval of the authority. If re-entry is inevitable, it is important that the actions of all concerned departments are well co-ordinated and the other departments are all fully informed of any further development of the situation since the determination of the Contractor's employment in one contract is likely to affect all other contracts between the same Contractor and the Government of the HKSAR. The relevant documents to follow under this Clause are WBTC No. 16/99 with amendments A, B and C, and No. 11/2001.

Analysis and application

Clause 81(1) "If the Contractor shall become bankrupt ... or (being a corporation) shall go into liquidation ..."

Bankruptcy or liquidation could entitle the Employer to re-enter the Site but there may be a case for the trustee or liquidator to arrange novation of the Contract using a new contractor who is qualified and acceptable to take over the Works. If there is no such agreement or arrangement, bankruptcy or liquidation will usually result in other breach of the Contract, such as excessive delay or the Contractor persistently in breach of his obligation under the Contract, and the Employer can also re-enter the Site on such other grounds.

Clause 81(1) "... or if the Contractor shall assign the Contract without the consent in writing of the Employer ... or shall have an execution levied on his goods ..."

These will breach Clause 3 or Clauses 71 and 72, entitling the Employer to re-enter but such situations seldom occur.

Clause 81(1) "... or if the Engineer shall certify in writing to the Employer that in his opinion the Contractor: (a) has abandoned the Contract, or ... (e) despite previous warning by the Engineer in writing is failing to proceed with the Works with due diligence or is persistently in breach of any of his obligations under the Contract, or ... (g) has to the detriment of good workmanship or in defiance of the Engineer's instruction to the contrary sub-contracted any part of the Works, ..."

The most common cause relied on by the Employer to re-enter is item (e) when the Contractor fails to proceed with the Works with due diligence causing excessive delay to the Contract or in serious breach of his obligations under the Contract. To satisfy this criterion, the Engineer must have given sufficient and repetitive warnings to the Contractor. It should be noted that failing to proceed with due diligence would cause excessive delay to the Works, and the Employer's loss may not be fully recovered from the liquidated damages and the Contractor would not be able to contend that he should remain on site to complete the Works provided that he can pay the liquidated damages.[3] See also the commentary to Clause 82.

In this major viaduct contract, if the original Contractor has got bankrupt and the Government's maintenance contractor has been employed to complete all the other outstanding works, the latter has completed all the surfacing and now almost $10M is due to the Employer by the Contractor. The Contractor's Plant has a resale value of $8M. Can the Employer now sell all the Contractor's Plant and use the money to pay the maintenance contractor?

The HKSAR Conditions:

Determination of the Contractor's Employment

(continued)

(1) *(Continued from p. 198)* then the Employer may after giving at least 7 days' notice in writing to the Contractor enter upon the Site and the Works and expel the Contractor therefrom without thereby avoiding the Contract or releasing the Contractor from any of his obligations or liabilities under the Contract or affecting the rights and powers conferred on the Employer or the Engineer by the Contract and the Employer may complete the Works or may employ any other contractor to complete the Works and the Employer or such other contractor may use for such completion so much of the Constructional Plant, temporary buildings and materials which become the property of the Employer under Clauses 71 and 72 as the Employer may think proper and the Employer may at any time sell any of the said Constructional Plant, temporary buildings and unused materials and apply the proceeds of sale in or towards the satisfaction of any sum due or which may become due to the Employer from the Contractor under the Contract.

(2) As soon as may be practicable after such entry and expulsion by the Employer, the Engineer shall ascertain and record:

(a) the quantity of work completed up to the time of such entry and expulsion, and

(b) the quantity of unused or partially used materials and list any Constructional Plant and temporary buildings which have become the property of the Employer under the Contract as at the time of such entry and expulsion.

The provisions of Clause 59 shall apply for the purposes of attendance by the Contractor for measurement and agreement of records and drawings.

Corresponding clauses in the ICE Conditions:

Clause 65 *(continued)*

(1) *(Continued from p. 198)* then the Employer may after giving 7 days' notice in writing to the Contractor specifying the event relied on enter upon the Works and any other parts of the Site provided by the Employer and expel the Contractor therefrom without thereby avoiding the Contract or releasing the Contractor from any of his obligations or liabilities under the Contract.

Where a notice of termination is given pursuant to a certificate issued by the Engineer under this sub-clause it shall be given as soon as is reasonably possible after receipt of the certificate.

Provided that the Employer may extend the period of notice to give the Contractor an opportunity to remedy the situation.

(2) Where the Employer has entered upon the Works and any other parts of the Site as set out in sub-clause (1) of this Clause he may

(a) complete the Works himself or

(b) employ any other contractor to complete the Works

and in either case may use for such completion any of the Contractor's Equipment Temporary Works goods and materials on any part of the Site.

The Employer may at any time sell any of the said Contractor's Equipment Temporary Works and unused goods and materials and apply the proceeds of sale in or towards the satisfaction of any sums due or which may become due to him from the Contractor under the Contract.

(4) As soon as may be practicable after any such entry and expulsion by the Employer the Engineer shall fix and determine as at the time of such entry and expulsion

(a) the amount (if any) which has been reasonably earned by or would reasonably accrue to the Contractor in respect of work actually done by him under the Contract and *(to be continued on page 206)*

Corresponding clauses in the FIDIC Conditions:

Clause 15.2 (*continued*)

(*Continued from p. 199*) In any of these events or circumstances, the Employer may, upon giving 14 days' notice to the Contractor, terminate the Contract and expel the Contractor from the Site. However, in the case of sub-paragraph (e) or (f), the Employer may by notice terminate the Contract immediately.

The Employer's election to terminate the Contract shall not prejudice any other rights of the Employer, under the Contract or otherwise.

The Contractor shall then leave the Site and deliver any required Goods, all Contractor's Documents, and other design documents made by or for him, to the Engineer. However, the Contractor shall use his best efforts to comply immediately with any reasonable instructions included in the notice (i) for the assignment of any subcontract, and (ii) for the protection of life or property or for the safety of the Works.

After termination, the Employer may complete the Works and/or arrange for any other entities to do so. The Employer and these entities may then use any Goods, Contractor's Documents and other design documents made by or on behalf of the Contractor.

The Employer shall then give notice that the Contractor's Equipment and Temporary Works will be release to the Contractor at or near the Site. The Contractor shall promptly arrange their removal, at the risk and cost of the Contractor. However, if by this time the Contractor has failed to make a payment due to the Employer, these items may be sold by the Employer in order to recover this payment. Any balance of the proceeds shall then be paid to the Contractor.

Clause 15.3

As soon as practicable after a notice of termination under Sub-Clause 15.2 [*Termination by Employer*] has taken effect, the Engineer shall proceed in accordance with Sub-Clause 3.5 [*Determinations*] to agree or determine the value of the Works, Goods and Contractor's Documents, and any other sums due to the Contractor for work executed in accordance with the Contract.

條款 81

香港特別行政區土木工程合同一般條件：

終止僱用「承建商」

（續）

(1) (續頁199) 則「僱主」可以在給予「承建商」最少7天書面通知後進駐「工地」及「本工程」並將「承建商」逐出「工地」和「本工程」，而這並不使「合同」無效或免除「承建商」在「合同」內的任何責任或法律責任或影響「合同」給予「僱主」或「工程師」的權利及權力，「僱主」也可自行或僱用任何其他承建商去完成「本工程」，且「僱主」或此等承建商為完成「本工程」可在「僱主」認為適當的情況下使用根據第71及72條規定已成為「僱主」財產的「施工設備」、「臨時建築物」及料物，「僱主」亦可隨時出售任何此等「施工設備」、「臨時建築物」及剩餘的物料並將出售所得收益用來支付按照「合同」「承建商」到期須支付「僱主」或可能須支付「僱主」的任何金額。

(2) 「僱主」在如此進駐「工地」及「本工程」並將「承建商」逐出後，「工程師」須在實際可行的情況下盡快確實及記錄：

   (a) 直至如此進駐「工地」及「本工程」並將「承建商」移出的時間為止，「承建商」已完成的工程的數量；及

   (b) 在進駐「工地」及「本工程」並將「承建商」逐出時所剩餘或部分已使用的物料數量，並列出按照「合同」已成為「僱主」財產的任何「施工設備」及「臨時建築物」。

   第59條的條文須用於「承建商」出席為進行計量及同意紀錄及圖則的事宜。

Analysis and application

Clause 81(1) "… the Employer may after giving at least 7 days' notice in writing to the Contractor enter upon the Site and the Works and expel the Contractor therefrom without thereby avoiding the Contract or releasing the Contractor from any of his obligations …"

The 7 days' notice in writing may be a problem because the Contractor might try to remove most of his valuable Plant or materials or other valuable things from the Site. The Engineer must rely on the relevant HKSAR Clause on Vesting of Construction Plant and temporary buildings and materials to prohibit their removal. As explained before in the commentary, the provision merely tells the Contractor to leave the Site and the Employer takes over. The Contractor's obligations still continue where applicable and the Employer's right and the Engineer's power are still retained.

Clause 81(1) "… the Employer may complete the Works or may employ any other contractor to complete the Works …"

What the Employer should do depends on the state of completion of the Works. If the Works are close to completion, the Government of the HKSAR can use his own maintenance contractor first to protect the Works and implement any safety measures, such as lighting, guarding and signing, and then complete any outstanding works. If there is still a very substantial part of the Works to be completed, the Government may call a new Contract, usually following open Tender procedures, for the completion of the remaining Works. Nevertheless, the maintenance contractor still needs to be deployed to take over the Works for temporary protection and carry on any critical activities before the new Contractor arrives.

Clause 81(1) "… the Employer or such other contractor may use for such completion so much of the Constructional Plant, temporary buildings and materials … as the Employer may think proper …"

The Engineer must first make a complete record of the Constructional Plant and materials on Site, together with good record photographs before using the Constructional Plant and materials. It would be necessary to ensure that such Plant or materials belongs to the Contractor and that all necessary tests for suitability of operation or incorporation into the Works are carried out. Where the Plant is to be reused to complete the Works, the Employer should include a disclaimer clause into the new Contract. Special attention has to be paid to any hired or hired-purchase Plant. The Engineer must, in any case, have obtained details of any hired or hired-purchase agreement. He must consider whether it could be advantageous to have the benefit of their agreements assigned to the Employer.

Clause 81(1) "… the Employer may at any time sell any of the said Constructional Plant, temporary buildings and unused materials and apply the proceeds of sale towards the satisfaction of any sum due …"

Any sale should not be initiated unless it can be demonstrated by the usual accounting procedure that a debt exists. For the sale of Constructional Plant, consideration should take into account the condition of the Plant, the risks, the cost and time consequence of failure of the Plant if reused.

Clause 81(2) "As soon as may be practicable after such entry … the Engineer shall ascertain and record: (a) the quantity of work completed up to the time of such entry and expulsion, and (b) the quantity of unused or partially used material and list any Constructional Plant and temporary buildings which have become the property of the Employer under the Contract as at the time of such entry and expulsion.…"

The "quantity of work completed" should mean the quantity of permanent work which are to be re-measured, and the "quantity of unused or partially used materials" means those materials to be incorporated into the Works under Clause 72. Constructional Plant and temporary buildings, which have become the property of the Employer under the Contract are those vested under Clause 71. All these must be duly ascertained and recorded as soon as practical after re-entry.

 Q 49

There are only two more floors to construct to complete the fabric of the buildings. The Contractor suddenly goes bankrupt. What course of action can the Employer take?

Clause 81
Determination of the Contractor's Employment

(continued)

(3) By the notice referred to in sub-clause (1) of this Clause or by further notice in writing within 28 days of the date thereof the Employer may require the Contractor to assign to the Employer and if so required the Contractor shall forthwith assign to the Employer the benefit of any agreement for the supply of any materials and/or for the execution of any work for the purposes of this Contract which the Contractor may have entered into.

(4) If the Employer shall enter and expel the Contractor in accordance with this Clause, the Employer shall not be liable to pay to the Contractor any money on account of the Contract until the expiry of the Maintenance Period or, where there is more than one such Period, until the expiry of the latest Period and thereafter until the cost of completion and maintenance, damages for delay in completion (if any) and all other expenses incurred by the Employer have been ascertained and the amount thereof certified by the Engineer.

(5) The Contractor shall then be entitled to receive only such sum (if any) as the Engineer may certify would have been payable to the Contractor upon due completion by him less the amount certified for the purposes of sub-clause (4) of this Clause. If the amount certified in accordance with sub-clause (4) of this Clause shall exceed the sum which would have been payable to the Contractor upon due completion by him then the Contractor shall upon demand pay to the Employer the amount of such excess.

Corresponding clauses in the ICE Conditions:

Clause 65 (*continued*)

(Continued from p. 202) (b) the value of any unused or partially used goods and materials which are under the control of the Employer and shall certify accordingly.

The said determination may be carried out *ex parte* or by or after reference to the parties or after such investigation or enquiry as the Engineer may think fit to make or institute.

(3) Where the Employer has entered upon the Works and any other parts of the Site as hereinbefore provided the Contractor shall if so instructed by the Engineer in writing within 7 days of such entry assign to the Employer the benefit of any agreement which the Contractor may have entered into for the supply of any goods or materials and/or for the carrying out of any work for the purposes of the Contract.

(5) (a) If the Employer enters and expels the Contractor under this Clause he shall not be liable to pay the Contractor any money under the Contract (whether in respect of amounts certified by the Engineer or otherwise) unless or until the Engineer certifies that an amount is due to the Contractor under sub-clause (b).

(b) The Engineer shall certify the difference between

(i) such sum as would have been due to the Contractor if he had completed the Works together with any proceeds of the sale under sub-clause 2 of this Clause and

(ii) the costs of completing the Works (whether or not the Works are completed under a separate contract) damages for delay (if any) and all other expenses properly incurred by the Employer.

(c) Such difference as is certified by the Engineer in sub-clause (b) shall be a debt due to the Employer or Contractor as the case may be.

(d) If the Engineer is satisfied at any time prior to the completion of the works that such sum as calculated under sub-clause (b)(ii) exceeds such sum as calculated under sub-clause (b)(i) he may issue an interim certificate to that effect notwithstanding that the Works have not been completed and such interim certificate shall be considered a debt due from the Contractor to the Employer. Every certificate issued by the Engineer pursuant to this Clause shall be sent to the Employer and at the same time copied to the Contractor with such detailed explanation as may be necessary.

Corresponding clauses in the FIDIC Conditions:

Clause 15.4

After a notice of termination under Sub-Clause 15.2 [*Termination by Employer*] has taken effect, the Employer may:

(a) proceed in accordance with Sub-Clause 2.5 [*Employer's Claims*],

(b) withhold further payments to the Contractor until the costs of execution, completion and remedying of any defects, damages for delay in completion (if any),

and all other costs incurred by the Employer, have been established, and/or

(c) recover from the Contractor any losses and damages incurred by the Employer and any extra costs of completing the Works, after allowing for any sum due to the Contractor under Sub-Clause 15.3 [*Valuation at Date of Termination*]. After recovering any such losses, damages and extra costs, the Employer shall pay any balance to the Contractor.

條款 81 （續）

香港特別行政區土木工程合同一般條件：

終止僱用「承建商」

(3) 「僱主」憑着發出本條第(1)款提及的通知或在這通知發出日期後28天內再次發出的書面通知要求「承建商」將任何為本「合同」有關而訂立的任何供應物料及/或實施工程的協議的利益轉讓給「僱主」。如「僱主」提出要求，「承建商」須立刻將此等利益轉讓給「僱主」。

(4) 如「僱主」按照本條進駐並將「承建商」逐出「工地」及「本工程」，在保養期（如多於一個保養期，則指最後的一個）完結之前，「僱主」無法律責任為本「合同」向「承建商」支付任何款項；及其後在經「工程師」確定並經他核實的竣工及保養的費用，竣工延遲的賠償（如有的話）及「僱主」因此引致的所有其他開支之前，「僱主」亦無法律責任為本「合同」向「承建商」支付任何款項。

(5) 「承建商」則只有權收取經「工程師」核實妥善竣工時「僱主」須付給「承建商」的金額減除根據本條第(4)款核實的款項的餘額（如有的話）。如根據本條第(4)款核實的款項超出「承建商」妥善地竣工時須獲得支付的金額，則「承建商」必須因應「僱主」的要求向「僱主」支付此多出的款項。

Analysis and application

Clause 81(3) "By the notice referred to in sub-clause (1) of this Clause or by further notice in writing within 28 days … the Employer may required … and … the Contractor shall forthwith assign to the Employer the benefit of any agreement for the supply of any materials and/or for the execution of any work …"

It is desirable to retain the sub-contractors or suppliers and especially the Specialist Contractors or Suppliers in the ETWB Lists to carry out any outstanding works unless the sub-contractors or suppliers themselves are in default. The right to an assignment of the sub-contract will, however, not be effective unless the sub-contract permits assignment or the sub-contractor or suppliers themselves consent.[4] Early contact or negotiation with the sub-contractor or suppliers is essential because the sub-contractor or suppliers may also treat their contracts as having been abandoned by their employers for suspension lasting more than 90 days.

Clause 81(4) "… the Employer shall not be liable to pay the Contractor any money on account of the Contract until the expiry of the Maintenance Period … and thereafter until the cost of completion … and all other expenses incurred by the Employer have been ascertained …"

This sub-clause, together with sub-clause (5), provides that no payments are due from either party after determination until the expiry of the Maintenance Period and all the Employer's expenses have been ascertained and the amount certified by the Engineer. This will delay any recovery of money by the Employer unnecessarily since certification may not be made years after the determination. Therefore, the Government of the HKSAR now has a Special Condition superseding sub-clause (4) and (5) which provides that certification by the Engineer for any money due to either party may be made when the difference between the following can be ascertained:

(i) such sum as would have been due to the Contractor if he had completed the Works together with any proceeds of sale; and

(ii) the Cost of completing the Works, damages for delay and other expenses properly incurred by the Employer.

Comparison with the ICE Conditions

The ICE sub-clause 65(1) is very much the same as the HKSAR Clause 81(1) and (2), except that it requires a notice of termination to be given as soon as is reasonably possible to the Contractor after the receipt of a certificate of default (items (e) to (j)) from the Engineer. The Employer, however, may extend the period of notice to give the Contractor an opportunity to remedy the situation. The ICE Clause also provides for assignment of benefit of any agreement, which the Contractor entered into for the supply of goods, materials and/or for the carrying out of work for the purpose of the Contract. The ICE sub-clause (4) expressly allows for determination of the amount due to be carried out *ex parte* by the Engineer in case the Contractor is no longer available at the time of determination. The ICE sub-clause (5) allows for the Engineer's certification at any time (and not until the expiry of the Maintenance Period as in case of the original HKSAR Conditions) once the sum due to the Contractor or a debt due to the Employer can be ascertained after completing the Works.

Comparison with the the FIDIC Conditions

The FIDIC Conditions is similar to the HKSAR Conditions and the ICE Conditions. However, offering bribes or any other favour is also a cause for termination. It also expressly requires the Contractor to use his best efforts to comply immediately with any reasonable instructions included in the notice of re-entry for the protection of life or property or for the safety of the Works. There is no provision for the Employer to use the Contractor's Equipment and Temporary Works but these will be released to the Contractor who shall promptly arrange for their removal. For recovering money as a result of termination, the Employer can follow the procedures for Employer's Claim under sub-clause 2.5 or the procedures for valuation at the Date of Termination under sub-clause 15.3.

Notes

1. See the commentary to Clause 63(1) in Max Abrahamson, *Engineering Law and the ICE Contracts*, 4th Edition, p. 277, n. 1.

2. See the commentary to Clause 63(3) in Max Abrahamson, *Engineering Law and the ICE Contracts*, 4th Edition, p. 281, n. 3.

3. *Bath and North East Somerset District Council v Mowlem plc*. [2004] BLR153 CA.

4. See the commentary to Clause 63(2) in Max Abrahamson, *Engineering Law and the ICE Contracts*, 4th Edition, p. 280, n. 13.

There are some outstanding road surfacing works included in the Drawings but are outside the Site limit of the tunnel Contract. Can the Contractor refuse to carry out such works?

Clause 82

Work by Person Other than the Contractor

(1) If the Contractor shall fail to carry out any work required under the Contract or refuse to comply with any instruction or order given by the Engineer in accordance with the Contract within a reasonable time, the Engineer may give the Contractor 14 days' notice in writing to carry out such work or comply with such instruction. If the Contractor fails to comply with such notice, the Employer shall be entitled to carry out such work or instruction by his own workers or by other contractors. Without prejudice to any other remedy, all additional expenditure properly incurred by the Employer in having such work or instruction carried out shall be recoverable by the Employer from the Contractor.

(2) If by reason of any accident or failure or other event occurring to, in, or in connection with the Works any remedial or other work shall in the opinion of the Engineer be urgently necessary and the Contractor is unable or unwilling at once to do such remedial or other work, the Engineer may authorize the carrying out of such remedial or other work by a person other than the Contractor. If the remedial or other work so authorized by the Engineer is work which in the Engineer's opinion the Contractor was liable to do under the Contract, all expenses properly incurred in carrying out the same shall be recoverable by the Employer from the Contractor. Provided that the Engineer shall as soon after the occurrence of any such emergency as may be reasonably practicable notify the Contractor thereof in writing.

Corresponding clauses in the ICE Conditions:

Clause 62

If in the opinion of the Engineer any remedial or other work or repair is urgently necessary by reason of any accident or failure or other event occurring to or in connection with the Works or any part thereof either during the carrying out of the Works or during the Defects Correction Period the Engineer shall so inform the Contractor with confirmation in writing.

Thereafter if the Contractor is unable or unwilling to carry out such work or repair at once the Employer may himself carry out the said work or repair using his own or other workpeople.

If the work or repair so carried out by the Employer is work which in the opinion of the Engineer the Contractor was liable to carry out at his own expense under the Contract all costs and charges properly incurred by the Employer in so doing shall on demand be paid by the Contractor to the Employer or may be deducted by the Employer from any monies due or which may become due to the Contractor.

Corresponding clauses in the FIDIC Conditions:

Clause 11.4

If the Contractor fails to remedy any defect or damage within a reasonable time, a date may be fixed by (or on behalf of) the Employer, on or by which the defect or damage is to be remedied. The Contractor shall be given reasonable notice of this date.

If the Contractor fails to remedy the defect or damage by this notified date and this remedial Work was to be executed at the cost of the Contractor under Sub-Clause 11.2 [*Cost of Remedying Defects*], the Employer may (at his option):

(a) Carry out the work himself or by others, in a reasonable manner and at the Contractor's cost, but the Contractor shall have no responsibility for this work; and the Contractor shall subject to Sub-Clause 2.5 [*Employer's*

Claims] pay to the Employer the costs reasonably incurred by the Employer in remedying the defect or damage;

(b) Require the Engineer to agree or determine a reasonable reduction in the Contract Price in accordance with Sub-Clause 3.5 [*Determination*]; or

(c) If the defect or damage deprives the Employer of substantially the whole benefit of the Works or any major part of the Works, terminate the Contract as a whole, or in respect of such major part which cannot be put to the intended use. Without prejudice to any other rights, under the Contract or otherwise, the Employer shall then be entitled to recover all sums paid for the Works or for such part (as the case may be), plus financing costs and the cost of dismantling the same, clearing the Site and returning Plant and Materials to the Contractor.

條款 82

由「承建商」以外人士施行工程

(1) 如「承建商」未能在合理的時間內施行按照「合同」規定的任何工程或拒絕遵照「工程師」根據「合同」發出的任何指令或命令，「工程師」可以書面通知「承建商」，要求他在14天內施行此等工程或遵從此等指令。如「承建商」未能遵從此通知，「僱主」有權用他自己的工人或其他承建商進行該等工程或指令。在不損害任何其他補救措施下，「僱主」可向「承建商」追討因施行此等工程或指令而恰當地招致的所有額外支出。

(2) 如因任何意外或故障或其他事故發生於「本工程」或與「本工程」有關而使「工程師」認為有急切需要進行任何補救工程或其他工程，但「承建商」未能或不願立刻進行此等補救工程或其他工程，「工程師」可授權「承建商」以外的人進行此等補救工程或其他工程。如「工程師」認為如此授權施行的補救工程或其他工程按照「合同」是「承建商」有法律責任進行的，「僱主」可向「承建商」追討因進行此等工程而恰當地招致的所有支出。但「工程師」必須在此等緊急事故發生後切實合理的時間內以書面通知「承建商」。

Commentary

This Clause consists of two parts. The first part empowers the Employer to use his own workers or other contractors to carry out the work stipulated under the Contract or instructed by the Engineer, which the Contractor fails to carry out. In this case, the power is invoked after a 14 days' notice is given by the Engineer to the Contractor and the Cost of the work can be recoverable by the Employer. The second part is in connection with the Contractor's failure to carry out remedial works or other works of urgent nature. In such case, the Engineer is immediately empowered to authorize the remedial or urgent works to be carried out by other persons. If such work is decided by the Engineer to be the responsibility of the Contractor, the Cost can be recoverable by the Employer. Money can be deducted from payments due to the Contractor under this Contract or other contracts with the Government under Clause 83. The application of this Clause is not limited to the Contract Period but also extends to the Maintenance Period and the Clause overlaps with Clause 56, sub-clause (4) "Execution of Work of Repair" and Clause 58, sub-clause (1) "Investigating Defects." Furthermore, this Clause may be invoked for the convenience of carrying out any urgent protection work or outstanding work when the Contractor abandons the Contract or when he goes to liquidation.

Analysis and application

Clause 82(1) "… or refuse to comply with any instruction or order given by the Engineer in accordance with the Contract …"

Clause 15 stipulated that the Contractor shall adhere strictly to the Engineer's instruction on any matter related to the Contract whether mentioned in the Contract or not. Clause 22, sub-clause (2) requires the Contractor to make good any damage, which may occur to any property of the Employer. Hence, the application of this Clause 82 is very wide and is not restricted to the "Works" included in the Contract. For example, the Contractor may be liable for any damage to any existing street furniture, noise barrier, fencing, etc. and he may be instructed to carry out the repair or replacement work, and if he fails to do so, this Clause 82 can be invoked.

Clause 82(1) "… If the Contractor fails to comply with such notice, the Employer shall be entitled to carry out such work or instruction by his own workers or by other contractors…"

If the Contractor delays beyond the due date for completion, the Employer may wish to complete the Works himself to enable occupation and use by the Employer, but the Contractor refuses to do so nor gives access to the Employer's own workers to carry out the Works, and the Contractor insists that he should do the work himself and he is willing to pay the liquidated damages. The Employer may seek an injunction to restrain the Contractor from barring entry to the Site since the Court decided in one case that the liquidated damages may have under-estimated the Employer's true loss.[1]

Clause 82(1) "… Without prejudice to any other remedy, all additional expenditure properly incurred by the Employer … shall be recoverable by the Employer from the Contractor."

The wording "all additional expenditure" possibly means the difference between the expenditure properly incurred in carrying out the work by the Employer's own workers or contractors, including any administrative costs, and the cost if the work is carried out by the Contractor based on the rates in the Bill of Quantities. The dilemma the Government of the HKSAR often faces is that the work may be cheaper if carried out by the Government's contractor, even including any further administrative cost in arranging the work by the Government officers and then there will be no "additional expenditure incurred." Should we pay the Contractor the work he failed to carry out in accordance with the Bill of Quantities rates and let the Contractor have a windfall? It appears that the question may be academic only for the following reasons. Firstly, the Contractor is already in breach of his obligation and the Employer is entitled to other remedy because of the wording "Without prejudice to any other remedy." It is, therefore, submitted that the Employer has an option not to pay the Contractor in accordance with the Bill rates for the works he fails to carry out. Secondly, the Contractor seldom deliberately refuses to carry out the Contract Works simply because of under-pricing since he would take the risk of being given an adverse performance report. Thirdly, if the works concerned are not large in quantity, the administrative cost incurred in arranging the works, involving both the consultants and the Government officers could be very substantial and in many cases would exceed the difference between the Contract prices and maintenance contract prices.

Clause 82(2) "… the Contractor is unable or unwilling at once to do such remedial or other work, the Engineer may authorize the carrying out of such remedial or other work by a person other than the Contractor.…"

For the remedial work or works of an urgent nature, the Engineer can order the carrying out of such work by another person. There is no requirement to give notice in advance although a short notice is still desirable if time permits. As regards recovery by the Employer, there is no specific mentioning of how the "additional expenditure" should be calculated and hence the recovery is effected by calculating the damages for breach of Contract.[2]

Clause 82(2) "… Provided that the Engineer shall as soon after the occurrence of any such emergency as may be reasonably practicable notify the Contractor thereof in writing."

During the carrying out of urgent works by another person, the Engineer must notify the Contractor to avoid any misunderstanding or the Contractor mistakenly incurs expenditure in such urgent works.

Comparison with the ICE Conditions

Clause 62 of the ICE Conditions is similar to sub-clause 82(2) of the HKSAR Conditions for the right to employ others to carry out remedial work urgently necessary and there is no equivalent to sub-clause 82(1). The ICE Clause expressly mentions its applicability during the Defects Correction Period.

Comparison with the FIDIC Conditions

The FIDIC Clause 11.4 is wider than the ICE Clause 62 in that it is applicable to remedial work, which the Contractor fails to rectify within a reasonable time fixed by the Employer. Sub-Clause (a) gives the Employer the option to carry out the remedial works himself, but expressly states that the Contractor shall not be responsible for the remedial works. The cost "reasonably incurred by the Employer" shall be recoverable from the Contractor as the Employer's Claims pursuant to Sub-Clause 2.5. Sub-Clause (b) empowers the Engineer to agree with the Contractor or determine a "reasonable reduction" in the Contract Price in accordance with Sub-Clause 3.5.

Notes

1. *Bath and North East Somerset District Council v Mowlem plc.* [2004] BLR153 CA
2. See the commentary to Clause 62 in Max Abrahamson, *Engineering Law and the ICE Contracts*, by 4th Edition, p. 275, n. 5.

Q 51 This is the last section of the cut and cover tunnel but the Contractor is unwilling to commit more expenditure to complete the Works on time but relies on his hundreds of unjustified claims for money and extension on time (EOT) to drag on. What should the Engineer do?

Clause 83

Recovery of Money Due to the Employer

(1) All damages (including liquidated damages), costs, charges, expenses, debts or sums for which the Contractor is liable to the Employer under any provision of the Contract may be deducted by the Employer from monies due to the Contractor under the Contract including Retention Money and the Employer shall have the power to recover any balance not so deducted from monies due to the Contractor under any other contract between the Employer and the Contractor.

(2) All damages (including liquidated damages), costs, charges, expenses, debts or sums for which the Contractor is liable to the Employer under any provision of any other contract between the Contractor and the Employer may be deducted by the Employer from monies due to the Contractor under the Contract, including Retention Money.

Corresponding clauses in the ICE Conditions

There is no corresponding clause. But various Clauses, namely Clause 25(3) "Failure to Insure"; Clause 39(2) "Removal of Unsatisfactory Work and Material"; Clause 47(5)(a) "Recovery of Liquidated Damages"; Clause 49(4) "Failure to Carry Out the Works Required", give provisions to deduct money due to the Contractor.

Corresponding clauses in the FIDIC Conditions:

Clause 2.5

If the Employer considers himself to be entitled to any payment under any Clause of these Conditions or otherwise in connection with the Contract, and/or to any extension of the Defects Notification Period, the Employer or the Engineer shall give notice and particulars to the Contractor. However, notice is not required for payments due under Sub-Clause 4.19 [*Electricity, Water and Gas*], under Sub-Clause 4.20 [*Employer's Equipment and Free-Issue Material*], or for other services requested by the Contractor.

The notice shall be given as soon as practicable after the Employer became aware of the event or circumstances giving rise to the claim. A notice relating to any extension of the Defects Notification Period shall be given before the expiry of such period.

The particulars shall specify the Clause or other basis of the claim, and shall include substantiation of the amount and/or extension to which the Employer considers himself to be entitled in connection with the Contract. The Engineer shall then proceed in accordance with Sub-Clause 3.5 [*Determinations*] to agree or determine (i) the amount (if any) which the Employer is entitled to be paid by the Contractor, and/or (ii) the extension (if any) of the Defects Notification Period in accordance with Sub-Clause 11.3 [*Extension of Defects Notification Period*].

This amount may be included as a deduction in the Contract Price and Payment Certificates. The Employer shall only be entitled to set off against or make any deduction from an amount certified in a Payment Certificate, or to otherwise claim against the Contractor, in accordance with this Sub-Clause.

條款 83

追討須支付「僱主」的款項

(1) 「僱主」可從按照「合同」須支付「承建商」的款項中，包括「保留金」，扣除「承建商」按照「合同」內任何條文有法律責任向「僱主」支付的一切賠償（包括經算定的賠償）、費用、收費、開支、債項或金額。「僱主」亦有權從按照他與「承建商」簽訂的任何其他「合同」而須支付「承建商」的款項中追討未有如上述方式扣除的任何餘款。

(2) 「僱主」可從按照「合同」須支付「承建商」的款項中，包括「保留金」，扣除「承建商」按照他與「僱主」簽訂的任何其他合同的任何條文而有法律責任向「僱主」支付的一切賠償（包括經算定的賠償）、費用、收費、開支、債項或金額。

Commentary

This Clause provides for the Employer to deduct any money owed to the Employer, e.g., liquidated damages or any payment for services rendered by the Employer from payment due to the Contractor or Retention Money, or if there is not enough payment due under the Contract, from payments due under other contracts between the Employer and the Contractor. It also provides for any money owed to the Employer arising from other contracts, to be deducted from the payment due or Retention Money under this Contract. While deduction from the payment due under the Contract itself is common, deduction from other contracts seldom takes place, except under the condition of re-entry, because the administrators of other contracts may not want the cash flows of other contracts being adversely affected or to sour the relationship with their respective contractors.

Analysis and application

Clause 83(1) " All damages … costs, … or sums for which the Contractor is liable to the Employer … may be deducted by the Employer from monies due …'

It is submitted that in case of bankruptcy of the Contractor or under re-entry condition, the Contract administrator should be extremely careful about any further money to be paid out to the Contractor under any interim certificate. He should urgently seek advice from the Financial Services Treasury Bureau and the Legal Advisory Division about the deduction he can make under his or other contracts, bearing in mind that any over-payment by the Government of the HKSAR as a whole may not be recovered since the money will go to the liquidator. For example, any liability of the Contractor on the damage of water mains, though still disputed, must be considered for deduction and quick advice should be sought from the Legal Advisory Division.

The rest of this Clause 83 is self-explanatory and straightforward.

Comparison with the ICE Conditions

The ICE Conditions has various clauses, namely, Clauses 25(3), 39(2), 47(5)(a) and 49(4), etc. which allows the Employer to deduct money from the payment due to the Contractor under this Contract. It does not expressly provide for deduction from other contracts with the Employer or deduction of money from this Contract arising from liability of the Contractor in other contracts with the Employer.

Comparison with the FIDIC Conditions

The FIDIC Clause 2.5 describes the procedure and entitlement of the Employer to pursue claims against the Contractor. In general, the Employer shall give notice and substantiate his claim, and try to agree the amount with the Contractor, failing which the Engineer shall make a determination. The agreed or determined amount can be deducted from the Payment Certificate.

Clause 86

Settlement of Disputes

(1) If any dispute or difference of any kind whatsoever shall arise between the Employer and the Contractor in connection with or arising out of the Contract or the carrying out of the Works including any dispute as to any decision, instruction, order, direction, certificate or valuation by the Engineer whether during the progress of the Works or after their completion and whether before or after the termination, abandonment or breach of the Contract, it shall be referred to and settled by the Engineer who shall state his decision in writing and give notice of the same to the Employer and the Contractor. Unless the Contract shall have been already terminated or abandoned the Contractor shall in every case continue to proceed with the Works with all due diligence and he shall give effect forthwith to every such decision of the Engineer unless and until the same shall be revised in mediation or arbitration as hereinafter provided. Such decision shall be final and binding upon the Contractor and the Employer unless either of them shall require that the matter be referred to mediation or arbitration as hereinafter provided.

Corresponding clauses in the ICE Conditions:

Clause 66

(1) In order to overcome where possible the causes of disputes and in those cases where disputes are likely still to arise to facilitate their clear definition and early resolution (whether by agreement or otherwise) the following procedure shall apply for the avoidance and settlement of disputes.

(2) If at any time
(a) the Contractor is dissatisfied with any act or instruction of the Engineer's Representative or any other person responsible to the Engineer or
(b) the Employer or the Contractor is dissatisfied with any decision opinion instruction direction certificate or valuation of the Engineer or with any other matter arising under or in connection with the Contract or the carrying out of the Works

the matter of dissatisfaction shall be referred to the Engineer who shall notify his written decision to the Employer and the Contractor within one month of the reference to him.

(3) The Employer and the Contractor agree that no matter shall constitute nor be said to give rise to a dispute unless and until in respect of that matter
(a) the time for the giving of a decision by the Engineer on a matter of dissatisfaction under Clause 66(2) has expired or the decision given is unacceptable or has not been implemented and in consequence the

Employer or the Contractor has served on the other and on the Engineer a notice in writing (hereinafter called the Notice of Dispute)
(b) an adjudicator has given a decision on a dispute under Clause 66(6) and the Employer or the Contractor is not giving effect to the decision, and in consequence the other has served on him and the Engineer a Notice of Dispute

and the dispute shall be that stated in the Notice of Dispute. For the purposes of all matters arising under or in connection with the Contract or the carrying out of the Works the word "dispute" shall be construed accordingly and shall include any difference.

(4) (a) Notwithstanding the existence of a dispute following the service of a Notice under Clause 66(3) and unless the Contract has already been determined or abandoned the Employer and the Contractor shall continue to perform their obligations.
(b) The Employer and the Contractor shall give effect forthwith to every decision of
(i) the Engineer on a matter of dissatisfaction given under Clause 66(2) and
(ii) the adjudicator on a dispute given under Clause 66(6)

unless and until that decision is revised by agreement of the Employer and Contractor or pursuant to Clause 66.

Corresponding clauses in the FIDIC Conditions:

Clause 20.4

(*"Dispute Adjudication Board (DAB)" as it appears below is defined in Sub-Clause 1.1.2.9 in the FIDIC Conditions of Contract for Construction as meaning the person or three persons so named in the Contract, or other person(s) appointed under Sub-Clause 20.2 [Appointment of the Dispute Adjudication Board] or Sub-Clause 20.3 [Failure to Agree Dispute Adjudication Board]*)

If a dispute (of any kind whatsoever) arises between the Parties in connection with, or arising out of, the Contract or the execution of the Works, including any dispute as to any certificate, determination, instruction, opinion or valuation of the Engineer, either Party may refer the dispute in writing to the DAB for its decision, with copies to the other Party and the Engineer. Such reference shall state that it is given under this Sub-Clause.

For a DAB of three persons, the DAB shall be deemed to have received such reference on the date when it is received by the chairman of the DAB.

Both Parties shall promptly make available to the DAB all such additional information, further access to the Site, and appropriate facilities, as the DAB may require for the purposes of making a decision on such dispute. The DAB shall be deemed to be not acting as arbitrator(s).

Within 84 days after receiving such reference, or within such other period as may be proposed by the DAB and approved by both Parties, the DAB shall give its decision, which shall be reasoned and shall state that it is given under this Sub-Clause. The decision shall be binding on both Parties, who shall promptly give effect to it unless and until it shall be revised in an amicable settlement or an arbitral award as described below. Unless the Contract has already been abandoned, repudiated or terminated, the Contractor shall continue to proceed with the Works in accordance with the Contract.

條款 86

香港特別行政區土木工程合同一般條件：

糾紛的解決

(1) 如「僱主」與「承建商」關乎或由於「合同」或「本工程」的施行而產生無論是任何種類的糾紛或分歧，包括關於「工程師」的任何決定、指令、命令、指示、證明書或價值的釐訂的任何糾紛，無論在「本工程」施行期間或竣工後，也無論在「合同」被終止、放棄或違反之前或之後，此糾紛或分歧必須提交「工程師」解決，而「工程師」必須以書面說明其決定，並將此決定通知「僱主」及「承建商」。除「合同」已被終止或放棄外，「承建商」在任何情況下皆必須盡力繼續進行「本工程」，並立即執行「工程師」的此等每項決定，除非及直至此等決定按以下規定經調解或仲裁而被修改為止。此等決定是最終決定，對「承建商」及「僱主」皆有約束力，除非任何一方要求此等決定的事宜按以下規定提交調解或仲裁。

Commentary

As the title suggests, this Clause states the procedures to be followed for the settlement of any dispute or difference between the Employer and the Contractor in connection with or arising out of the Contract or the carrying out of the Works, including any decision, instruction, order, direction, certificate or valuation by the Engineer. As explained later in the commentary, the Clause has wide coverage regarding the matters falling within the definition of dispute or difference. The Clause covers referral to the Engineer for a decision (the Engineer's decision is often referred to as the "Clause 86 decision"). While the Engineer must be fair and impartial in reaching his decision, he is under no obligation to have a hearing similar to that of an arbitrator, since the Engineer is not bound by the rule of natural justice in his decision.[1] However, he must act independently[2] from any restraint imposed by the Employer, unless it is made known in the Contract between the Employer and the Contractor.[3] He is also well advised to give an opportunity for both the Contractor and the Employer to make their submission or otherwise he may be accused of being unfair.[4] If either Party is not satisfied, he could refer to mediation any time during the construction period or after the completion of the Works. If there is a refusal to mediate or the time limit to respond to mediation expires, either Party could refer the dispute to arbitration. For arbitration it cannot be commenced until after the completion of the Works unless there is agreement between the Employer and the Contractor. It should be noted that if the Employer refuses mediation, it would take a long time before the dispute can be referred to arbitration, i.e., after the completion of the Works and during that time the Contractor must give effect to the Engineer's decision. For the Works Departments of the Development Bureau, the decision to mediate lies with the Director of the Department, who shall consult the Department of Justice before making the decision and in most cases, the Department agrees to mediate unless the Contractor's case has very little merit and the issue has a clear cut answer. In the latter case, the Parties may choose a document only or a short form arbitration. The document only arbitration does not require a hearing while the short form arbitration may include a very brief hearing. The latter is a standard provision in the Architectural Services Department for disputes arising out of a contract incorporating the service of a Dispute Resolution Adviser. In comparison with the ICE Conditions and the FIDIC Conditions, an adjudicator or Dispute Adjudication Board may be called up by either Party to decide the dispute by adjudication while adjudication is not provided in this HKSAR Clause. However, a Special Condition is now incorporated in all the HKSAR contracts, which has a provision for voluntary adjudication.

Analysis and application

Clause 86(1) "If any dispute or difference of any kind whatsoever shall arise between the Employer and the Contractor in connection with or arising out of the Contract or the carrying out of the Works ..."

The wording "any dispute or difference of any kind ... in connection with or arising out ..." suggests that this arbitration Clause has a very wide scope which even covers tortious claims. However, each case has to be considered on its own merits. In the Government of the HKSAR, performance reports of the Contractor are written on a quarterly basis and if the Contractor is not satisfied with the performance gradings, he may appeal to an internal panel. Certain contractors attempted to treat the issue as a dispute but were not successful since it is an internal matter and is outside the jurisdiction of the Clause.

Clause 86(1) "... during the progress of the Works or after their completion and whether before or after the termination, abandonment or breach of the Contract, ..."

As far as timing is concerned, the claim is extremely wide and it starts from the commencement of the Works to after their completion and applies even after the termination or abandonment of the Contract.

Clause 86(1) "... it shall be referred to and settled by the Engineer ..."

As a condition precedent to the resolution procedure, the matter should first be decided by the Engineer, who must give his decision in writing. However, since mediation is a consensual dispute resolution process, it is up to the Parties to decide whether the Engineer's decision is necessary and this step is sometimes bypassed.

Clause 86(1) "... shall in every case continue to proceed with the Works with all due diligence and he shall give effect forthwith to every such decision of the Engineer ..."

The progress of the Works must be maintained diligently despite the Contractor raises a dispute and the Engineer's decision, even though unreasonable, must be complied with until his decision is revised in mediation or arbitration.

 The spacing of the struts is now half of that certified by the independent checking engineer because of the Engineer's conservative stand about the ground condition. Can the Contractor raise a dispute and refer to arbitration now?

Clause 86

Settlement of Disputes

(continued)

(1) If the Engineer shall fail to give such decision for a period of 28 days after being requested to do so or if either the Employer or the Contractor be dissatisfied with any such decision of the Engineer then either the Employer or the Contractor may within 28 days after receiving notice of such decision, or within 28 days after the expiry of the said decision period of 28 days, as the case may be, request that the matter be referred to mediation in accordance with and subject to *The Government of the Hong Kong Special Administrative Region Construction Mediation Rules* (the Mediation Rules) or any modification thereof being in force at the date of such request.

Corresponding clauses in the ICE Conditions:

Clause 66 *(continued)*

(5) (a) The Employer or the Contractor may at any time before service of a Notice to Refer to arbitration under Clause 66(9) by notice in writing seek the agreement of the other for the dispute to be considered under "The Institution of Civil Engineers' Conciliation Procedure 1999" or any amendment or modification thereof being in force at the date of such notice.

(b) If the other party agrees to this procedure any recommendation of the conciliator shall be deemed to have been accepted as finally determining the dispute by agreement so that the matter is no longer in dispute unless a Notice of Adjudication under Clause 66(6) or a Notice to Refer to arbitration under Clause 66(9) has been served in respect of that dispute not later than one month after receipt of the recommendation by the dissenting party.

(6) (a) The Employer and the Contractor each has the right to refer a dispute as to a matter under the Contract for adjudication and either party may give notice in writing (hereinafter called the Notice of Adjudication) to the other at any time of his intention so to do. The adjudication shall be conducted under "The Institution of Civil Engineers' Adjudication

Procedure 1997" or any amendment or modification thereof being in force at the time of the said Notice.

(b) Unless the adjudicator has already been appointed he is to be appointed by a timetable with the object of securing his appointment and referral of the dispute to him within 7 days of such notice.

(c) The adjudicator shall reach a decision within 28 days of referral or such longer period as is agreed by the parties after the dispute has been referred.

(d) The adjudicator may extend the period of 28 days by up to 14 days with the consent of the party by whom the dispute was referred.

(e) The adjudicator shall act impartially.

(f) The adjudicator may take the initiative in ascertaining the facts and the law.

(7) The decision of the adjudicator shall be binding until the dispute is finally determined by legal proceedings or by arbitration (if the contract provides for arbitration or the parties otherwise agree to arbitration) or by agreement.

(8) The adjudicator is not liable for anything done or omitted in the discharge or purported discharge of his functions as adjudicator unless the act or omission is in bad faith and any employee or agent of the adjudicator is similarly not liable.

Corresponding clauses in the FIDIC Conditions:

Clause 20.4 *(continued)*

If either Party is dissatisfied with the DAB's decision, then either Party may, within 28 days after receiving the decision, give notice to the other Party of its dissatisfaction. If the DAB fails to give its decision within the period of 84 days (or as otherwise approved) after receiving such reference, then either Party may, within 28 days after this period has expired, give notice to the other Party of its dissatisfaction.

In either event, this notice of dissatisfaction shall state that it is given under this Sub-Clause, and shall set out the matter in dispute and the reason(s) for dissatisfaction. Except as stated in Sub-Clause 20.7 [*Failure to Comply with Dispute Adjudication Board's Decision*] and Sub-Clause 20.8 [*Expiry of Dispute Adjudication Board's Appointment*], neither Party shall be entitled to commence arbitration of a dispute unless a notice of dissatisfaction has been given in accordance with this Sub-Clause.

If the DAB has given its decision as to a matter in dispute to both Parties, and no notice of dissatisfaction has been given by either Party within 28 days after it received the DAB's decision, then the decision shall become final and binding upon both Parties.

條款 86 （續）

香港特別行政區土木工程合同一般條件：

糾紛的解決

(1) 如「工程師」未能在被要求作出決定後28天內作出決定或如「僱主」或「承建商」對「工程師」的決定不滿意，「僱主」或「承建商」可在接獲此等決定的通知後28天內，或視情況而定，在上述28天決定期期滿後的28天內，要求將有關事宜按照及遵從《香港特別行政區政府建築業調解規則》(調解規則)及其在提出要求時的任何有效修改的規定下提交調解。

Analysis and application

Clause 86(1) "If the Engineer shall fail to give such decision for a period of 28 days after being requested to do so or if either the Employer or the Contractor be dissatisfied with any such decision of the Engineer … request that the matter be referred to mediation …"

The Engineer must first consider whether there is actually a dispute. It must be noted that for civil engineering claims, especially where extension of time and prolongation/disruption cost are involved, the matters in dispute (or the issues) are intermingled with one another. Unless a very detailed analysis is made on the Works programmes and their interface with other claims, the issues are not usually well defined. It is impractical to require the Engineer to give a decision within 28 days and there must be adequate time for the Contractor to submit proper claims and the Engineer to consider the claims. One practical way is for the Parties to extend the 28-day period until there is sufficient time for the Contractor and the Engineer to fulfil their respective duties in relation to the claims. Though not mentioned in the Clause itself, negotiation may be an alternative to mediation. For substantial claims, the project team as the Employer must obtain approval from the authority before commencing the negotiation. If the Parties agree to mediate, they should follow *The Government of the HKSAR Construction Mediation Rules* regarding the choice of mediator and the subsequent procedure to be followed. However, since mediation is a consensual process, the parties can adopt a procedure best suited to their own needs. In general, the following steps can be taken:

(i) Usually, the Contractor requests for mediation, suggesting one or a few mediators for agreement by the Government of the HKSAR.

(ii) The Government of the HKSAR, after consultation with the Legal Department, shall reply within 28 days.

(iii) Assuming there is an agreement to the mediator, each Party prepares his position paper setting out his position and arguments about the claim, and usually an independent expert and/or quantity surveyor are called to assist in establishing liability and/or quantum.

(iv) The mediator then calls a preliminary meeting to explain his role and subsequent procedure to follow and fix a timetable for exchange of documents. Each party also decides on the composition of his negotiating team. The need for legal representation depends on the nature of the dispute but it is usually not necessary.

(v) There will then be a mediation hearing, which usually lasts for a couple of days depending on the complexity of the issues. At the hearing, the mediator makes an opening statement, followed by each Party presenting his case. The mediator then directs the resolution process, taking each item in turn. The leader of each Party keeps making decisions about any concessions, usually based on a negotiating plan prepared in advance. The leader of each Party can call a stop to discuss with his team members privately at any time, with or without the presence of the mediator. The current tendency is that the Engineer can be present in the mediation as an observer and may assist the team leader when called upon to do so.

(vi) For the Government of the HKSAR, approval has to be obtained before the team leader commits any promise, and a provisional agreement subject to approval can be signed by the team leader.

Apart from following the Construction Mediation Rules and the above steps, the Government staff must follow the Administrative Guidelines in the *Works Bureau Technical Circular*.

Settlement of Disputes

Clause 86

(continued)

(2) If the matter cannot be resolved by mediation, or if either the Employer or the Contractor do not wish the matter to be referred to mediation then either the Employer or the Contractor may within the time specified herein require that the matter shall be referred to arbitration in accordance with and subject to the provisions of the *Arbitration Ordinance* (Cap. 341) or any statutory modification thereof for the time being in force and any such reference shall be deemed to be a submission to arbitration within the meaning of such Ordinance. Any reference to arbitration shall be made within 90 days of:

(a) the receipt of a request for mediation and subsequently the recipient of such request having failed to respond, or

(b) the refusal to mediate, or

(c) the failure of the mediation proceedings to produce a settlement acceptable to the Employer and the Contractor, or

(d) the abandonment of the mediation, or

(e) the Engineer failing to make a decision for a period of 90 days after being so requested to do so and subsequently neither the Employer nor the Contractor having requested mediation, or

(f) the receipt of a notice of a decision by the Engineer and subsequently neither the Employer nor the Contractor having requested mediation.

Corresponding clauses in the ICE Conditions:

Clause 66 *(continued)*

(9) (a) All disputes arising under or in connection with the Contract or the carrying out of the Works other than failure to give effect to a decision of an adjudicator shall be finally determined by reference to arbitration. The party seeking arbitration shall serve on the other party a notice in writing (called the Notice to Refer) to refer the dispute to arbitration.

(b) Where an adjudicator has given a decision under Clause 66(6) in respect of the particular dispute the Notice to Refer must be served within three months of the giving of the decision otherwise it shall be final as well as binding.

(10) (a) The arbitrator shall be a person appointed by agreement of the parties.

(b) If the parties fail to appoint an arbitrator within one month of either party serving on the other party a notice in writing (hereinafter call the Notice to Concur) to concur in the appointment of an arbitrator the dispute shall be referred to a person to be appointed on the application of either party by the President for the time being of the Institution of Civil Engineers.

(c) If an arbitrator declines the appointment or after appointment is removed by order of a competent court or is incapable of acting or dies and the parties do not within one month of the vacancy arising fill the vacancy then either party may apply to the President for the time being of the Institution of Civil Engineers to appoint another arbitrator to fill the vacancy.

In any case where the President for the time being of the Institution of Civil Engineers is not able to exercise the functions conferred on him by this Clause the said functions shall be exercised on his behalf by a Vice-President for the time being of the said Institution.

Corresponding clauses in the FIDIC Conditions:

Clause 20.6

Unless settled amicably, any dispute in respect of which the DAB's decision (if any) has not become final and binding shall be finally settled by international arbitration. Unless otherwise agreed by both Parties:

(a) the dispute shall be finally settled under the Rules of Arbitration of the International Chamber of Commerce,

(b) the dispute shall be settled by three arbitrators appointed in accordance with these Rules, and

(c) the arbitration shall be conducted in the language for communications defined in Sub-Clause 1.4 [*Law and Language*].

條款 **86** （續）	*香港特別行政區土木工程合同一般條件：* ## 糾紛的解決

(2) 如此事宜無法憑調解解決，或如果「僱主」或「承建商」不希望將有關事宜提交調解，「僱主」或「承建商」可在本條指定的時間內按照《仲裁條例》（第341章）或此條文文當時有效的任何法定修改及在其規定下要求將此事宜提交仲裁，而在任何此等情況下提交的仲裁必須被視為此條例所指的仲裁的提交。任何仲裁的提交必須於下列情況發生後90天內作出：

(a) 收到調解要求之後收件人沒有回應；或

(b) 拒絕調解；或

(c) 經調解程序後未能得到「僱主」及「承建商」皆接受的解決方案；或

(d) 放棄調解；或

(e) 「工程師」在被要求作出決定後90天內未能作出決定及其後「僱主」與「承建商」皆沒有要求調解；或

(f) 收到「工程師」所作決定的通知後「僱主」與「承建商」皆沒有重新要求調解。

Analysis and application

Clause 86(2) "If the matter cannot be resolved by mediation, or if either the Employer or the Contractor do not wish the matter to be referred to mediation then either the Employer or the Contractor may within the time specified herein require that the matter shall be referred to arbitration ..."

If any one Party does not want to mediate or if the mediation cannot produce any satisfactory results, either Party may within 90 days refer the matter to arbitration according to items (a) to (d) of this sub-clause (2). Though the matter is referred to arbitration, as explained in the commentary, the arbitration cannot be commenced until the completion of the Works unless there is agreement to arbitrate earlier by the Parties. Usually, the Government of the HKSAR would not agree. Items (e) and especially (f) are of great significance to the Contractor since after the Engineer has made a clear decision and the Contractor does nothing for the 90 days time limit, he will lose his right to bring the matter to any dispute resolution mechanism and the Engineer's decision will become final and binding upon the Contractor.

Clause 86

Settlement of Disputes

(continued)

(3) The arbitrator appointed shall have full power to open up, review and revise any decision (other than a decision under Clause 46(3) not to vary the Works), instruction, order, direction, certificate or valuation by the Engineer and neither Party shall be limited in the proceedings before such arbitrator to the evidence or arguments put before the Engineer for the purpose of obtaining his decision above referred to. Save as provided for in sub-clause (4) of this Clause no steps shall be taken in the reference to the arbitrator until after the completion or alleged completion of the Works unless with the written consent of the Employer and the Contractor.
Provided that:

(a) the giving of a certificate of completion in accordance with Clause 53 shall not be a condition precedent to the taking of any step in such reference;

(b) no decision given by the Engineer in accordance with the foregoing provisions shall disqualify him from being called as a witness and giving evidence before the arbitrator on any matter whatsoever relevant to the dispute or difference so referred to the arbitrator as aforesaid.

(4) In the case of any dispute or difference as to the exercise of the Engineer's powers under Clause 81(1) the reference to the arbitrator may proceed notwithstanding that the Works shall not then be or be alleged to be complete.

(5) The Hong Kong International Arbitration Centre Domestic Arbitration Rules shall apply to any arbitration instituted in accordance with this Clause unless the parties agree to the contrary.

(6) The reference to arbitration under sub-clause (2) of this Clause shall be a domestic arbitration for the purposes of Part II of the *Arbitration Ordinance* (Cap. 341).

Corresponding clauses in the ICE Conditions:

Clause 66 *(continued)*

(11) (a) Any reference to arbitration under this Clause shall be deemed to be a submission to arbitration within the meaning of the Arbitration Act 1996 or any statutory re-enactment or amendment thereof for the time being in force. The reference shall be conducted in accordance with the procedure set out in the Appendix to the Form of Tender or any amendment or modification thereof being in force at the time of the appointment of the arbitrator. Such arbitrator shall have full power to open up review and revise any decision opinion instruction direction certificate or valuation of the Engineer or an adjudicator.

(b) Neither party shall be limited in the arbitration to the evidence or arguments put to the Engineer or to any adjudicator pursuant to Clause 66(2) or 66(6) respectively.

(c) The award of the arbitrator shall be binding on all parties.

(d) Unless the parties otherwise agree in writing any reference to arbitration may proceed notwithstanding that the Works are not then complete or alleged to be complete.

(12) (a) No decision opinion instruction direction certificate or valuation given by the Engineer shall disqualify him from being called as a witness and giving evidence before a conciliator adjudicator or arbitrator on any matter whatsoever relevant to the dispute.

(b) All matters and information placed before a conciliator pursuant to a reference under sub-clause (5) of this Clause shall be deemed to be submitted to him without prejudice and the conciliator shall not be called witness by the parties or anyone claiming through them in connection with any adjudication arbitration or other legal proceedings arising out of or connected with any matter so referred to him.

Corresponding clauses in the FIDIC Conditions:

Clause 20.6 (*continued*)

The arbitrator(s) shall have full power to open up, review and revise any certificate, determination, instruction, opinion or valuation of the Engineer, and any decision of the DAB, relevant to the dispute. Nothing shall disqualify the Engineer from being called as a witness and giving evidence before the arbitrator(s) on any matter whatsoever relevant to the dispute.

Neither Party shall be limited in the proceedings before the arbitrator(s) to the evidence or arguments previously put before the DAB to obtain its decision, or to the reasons for dissatisfaction given in its notice of dissatisfaction. Any decision of the DAB shall be admissible in evidence in the arbitration.

Arbitration may be commenced prior to or after completion of the Works. The obligations of the Parties, the Engineer and the DAB shall not be altered by reason of any arbitration being conducted during the progress of the Works.

條款 86

（續）

香港特別行政區土木工程合同一般條件：

糾紛的解決

(3) 被委任的仲裁人有全權重新審議、覆查及修改「工程師」的任何決定（決定不根據第46條(3)款更改「本工程」除外）、指令、命令、指示、證明書或價值的釐訂。在仲裁人面前進行的程序中，任何一方所提出的證據或論據皆不限於為取得上述的「工程師」的決定曾提出的證據或論據。除本條第(4)款的規定外，任何一方不可在「承建商」完成或聲稱完成「本工程」之前就提交仲裁人仲裁的程序採取任何步驟，除非得到「僱主」及「承建商」書面同意。

但是：

(a) 「工程師」根據第53條頒發竣工證明書並不是採取任何提交仲裁程序步驟的先決條件；

(b) 「工程師」根據上述條文所作出的任何決定皆不會使他喪失被傳召為證人及在仲裁人面前就任何與提交給此仲裁人仲裁的糾紛或分歧有關的事宜作供的資格。

(4) 如果是關乎「工程師」根據第81條(1)款行使其權力的任何糾紛或分歧，雖然「承建商」並未完成或未聲稱完成「本工程」，提交仲裁人仲裁的程序仍可以進行。

(5) 《香港國際仲裁中心本地仲裁規則》適用於根據本條進行的任何仲裁，但雙方另有協議除外。

(6) 根據本條第(2)款提交的仲裁是《仲裁條例》（第341章）第II部所指的本地仲裁。

Analysis and application

Clause 86(3) "The arbitrator appointed shall have full power to open up, review and revise any decision (other than a decision under Clause 46(3) not to vary the Works), instruction, order, direction, certificate or valuation by the Engineer ..."

Apart from the Engineer's decision of not ordering a variation to deal with unsatisfactory work or materials in accordance with Clause 46(3), the arbitrator has very wide power to review and revise the Engineer's decisions, valuation or other instructions, etc. For example, if the Engineer has not ordered a variation, which the arbitrator finds that he should have made such an order, the arbitrator can rule that the Contractor is entitled to a "deemed variation."

Clause 86(3) "... neither party shall be limited in the proceedings before such arbitrator to the evidence or arguments put before the Engineer for the purpose of obtaining his decision ..."

Some contractors exploit their right to submit evidence and arguments not to the Engineer but only before the mediator or arbitrator and this could take the Employer by surprise. If this is the case, the Engineer must be called to assist and he must be given ample time to respond and consider the Contractor's further evidence and arguments. If fresh evidence is introduced at the hearing before the arbitrator, an adjournment of the hearing would be necessary for the Employer to consider his position and any further arguments.

Clause 86(3) "... no steps shall be taken in the reference to the arbitrator until after the completion or alleged completion of the Works unless with the written consent of the Employer and the Contractor...."

Arbitration proceeding can only be commenced after completion of the Works, unless both Parties agree, as explained earlier. This policy ensures that the Contractor continues with his work, though he disagrees with the decision of the Engineer.

Clause 86(4) "In the case of any dispute or difference as to the exercise of the Engineer's powers under Clause 81(1) the reference to the arbitrator may proceed notwithstanding that the Works shall not then be or be alleged to be complete."

The Engineer has the power to certify, for example, under item (a) of sub-clause 81(1) that the Contractor has abandoned the Contract and if the Contractor disputes his certificate, he can commence arbitration without waiting for the completion of the Works and this is an exception to sub-clause 86(3).

Clause 86(5) "The Hong Kong International Arbitration Centre Domestic Arbitration Rules shall apply to any arbitration instituted in accordance with this Clause ..."

The Rules describe essentially how to commence an arbitration, the appointment of the arbitrator, the holding of preliminary meeting, the exchange of statement of claims and statement of defense, the disclosure and inspection of documents, the representation, the hearing, the calling of witness, the power and jurisdiction of the arbitrator, the award and cost. A guideline entitled *A Guide to Arbitration under the Domestic Arbitration Rules 1993* published by the Hong Kong International Arbitration Centre gives practical directions in the use of the Rules in order to achieve maximum benefits from them.

Clause 86(6) "The reference to arbitration under sub-clause (2) of this Clause shall be a domestic arbitration for the purposes of Part II of *Arbitration Ordinance* (Cap. 341)."

For the Works contracts of the Government of the HKSAR, we are concerned with domestic arbitration only and not international arbitration.

Comparison with the ICE Conditions

Unlike the HKSAR Clause, the ICE Clause 66 gives a detailed procedure to effect a clear definition of the word "dispute." Before any dissatisfaction becomes a dispute, it must be referred to the Engineer who shall give a written decision to the Employer within one month. Before the matter is finally to be decided by arbitration, the Parties may voluntarily agree to adopt the conciliation procedure, which is similar to mediation in the HKSAR Clause. The recommendation of the conciliator will become final unless a Notice to Refer to arbitration is served by the dissenting Party within one month. If conciliation is not adopted, either Party may refer the dispute to adjudication and the decision of the adjudicator will be binding until the matter is finally decided by arbitration. For arbitration, unlike the HKSAR Clause, it can proceed before the completion of the Works.

Comparison with the FIDIC Conditions

If any dispute arises between the parties after the Engineer's determination in accordance with the FIDIC Clause 3.5, it is no longer necessary to refer to the Engineer for a decision. Either party may refer the dispute to the DAB in accordance with the FIDIC Clause 20.4, which consists of one person or three persons appointed before the commencement of Works on Site. The decision of DAB will be final unless either party gives a notice of dissatisfaction of the decision. Then either party can commence arbitration proceedings and the arbitration is international. There are three arbitrators, unless both Parties agree otherwise. Also, unlike the HKSAR Clause, either Party may commence arbitration before the completion of the Works.

Notes

1. *Amec Civil Engineering Ltd. v Secretary of State for Transport* [2005] BLR 227 CA.
2. *Scheldebouw v St. James Homes (Grosvenor Dock) Ltd. (2006)* EWHC89.
3. See also the commentary to Clause 2 explaining "referable decisions" on p. 12.
4. *Costain Ltd. v Bechtel* [2005] EWHC1018.

 When two Parties made an agreement with the provision for submitting their disputes to arbitration, does that mean that the Court would not be involved in any matters in the settlement of the disputes?

Clause 88

Default of the Employer

(1) In the event of the Employer failing to pay to the Contractor any sum certified in accordance with Clause 79 within 28 days after the same shall have become due under the provisions of the Contract the Contractor may give 14 days' notice in writing to the Employer to make payment of the sum due. Such notice shall make express reference to this Clause. In the event of failure by the Employer to make such payment within such 14 days notice period, the Contractor shall be entitled to terminate the Contract.

(2) So long as no notice pursuant to Clause 81(1) is given to the Contractor either before or during the 14 days' notice period provided in sub-clause (1) of this Clause, on expiration of that 14 days, the property in all Constructional Plant and temporary buildings brought upon the Site by the Contractor shall thereupon re-vest in him and he shall with all reasonable despatch remove the same from the Site.

(3) Nothing in this Clause shall prejudice the right of the Contractor to exercise, either in lieu of or in addition to the rights and remedies in this Clause specified, any other rights or remedies to which the Contractor may be entitled.

Corresponding clauses in the ICE Conditions:

Clause 64

(1) In the event that the Employer

(a) assigns or attempts to assign the Contract or any part thereof or any benefit or interest thereunder without the prior written consent of the Contractor or

(b) (i) becomes bankrupt or presents his petition in bankruptcy or

(ii) has a receiving order or administration order made against him or

(iii) makes an arrangement with or an assignment in favour of his creditors or

(iv) agrees to perform the Contract under a committee of inspection of his creditors or

(v) (being a corporation) has a receiver or administrator appointed or goes into liquidation (other than a voluntary liquidation for the purposes of amalgamation or reconstruction) or

(c) has an execution levied on his goods which is not stayed or discharged within 28 days

then the Contractor may after giving 7 days' notice in writing to the Employer specifying the event relied on terminate his employment under the Contract without thereby avoiding the Contract or releasing the Employer from any of his obligations or liabilities under the Contract provided that the Contractor may extend the period of notice to give the Employer an opportunity to remedy the situation.

(2) Upon expiry of the 7 days' notice referred to in sub-clause (1) of this Clause and notwithstanding the provisions of Clause 54 the Contractor shall with all reasonable dispatch remove from the site all Contractor's Equipment.

(3) Upon termination of the Contractor's employment pursuant to sub-clause (1) of this Clause the Employer shall be under the same obligations with regard to payment as if the Works had been abandoned under the provisions of Clause 63.

Provided that in addition to payments specified under Clause 63(4) the Employer shall pay to the Contractor the amount of any loss or damage to the Contractor arising from or as a consequence of such termination.

Corresponding clauses in the FIDIC Conditions:

Clause 16.2

The Contractor shall be entitled to terminate the Contract if:

(a) The Contractor does not receive the reasonable evidence within 42 days after giving notice under Sub-Clause 16.1 [*Contractor's Entitlement to Suspend Work*] in respect of a failure to comply with Sub-Clause 2.4 [*Employer's Financial Arrangements*],

(b) The Engineer fails, within 56 days after receiving a Statement and supporting documents, to issue the relevant Payment Certificate,

(c) The Contractor does not receive the amount due under an Interim Payment Certificate within 42 days after the expiry of the time stated in Sub-Clause 14.7 [*Payment*] within which payment is to be made (except for deductions in accordance with Sub-Clause 2.5 [*Employer's Claims*]),

(d) The Employer substantially fails to perform his obligations under the Contract,

(e) The Employer fails to comply with Sub-Clause 1.6 [*Contract Agreement*] or Sub-Clause 1.7 [*Assignment*],

(f) A prolonged suspension affects the whole of the Works as described in Sub-Clause 8.11 [*Prolonged Suspension*], or

(g) The Employer becomes bankrupt or insolvent, goes into liquidation, has a receiving or administration order made against him, compounds with his creditors, or carries on business under a receiver, trustee or manager for the benefit of his creditors, or if any act is done or event occurs which (under applicable Laws) has a similar effect to any of these acts or events.

In any of these events or circumstances, the Contractor may, upon giving 14 days' notice to the Employer, terminate the Contract. However, in the case of sub-paragraph (f) or (g), the Contractor may by notice terminate the Contract immediately.

條款 88

香港特別行政區土木工程合同一般條件：

「僱主」的失責

(1) 如果「僱主」沒有按照「合同」內的付款條文在到期28天內支付「承建商」任何根據第79條核實的金額，「承建商」可以給「僱主」14天的書面通知，要求「僱主」支付此項到期要付款的金額。此通知必須明確地提及本條。如「僱主」未能在這14天通知期內付款，「承建商」有權終止「合同」。

(2) 如在本條第(1)款規定的14天通知期內或之前，「承建商」沒有收到根據第81條(1)款發出的通知，則在這14天期限完結時，「承建商」攜帶入「工地」的所有「施工設備」及「臨時建築物」即重新歸「承建商」所擁有，「承建商」必須在合理時間內將該等「施工設備」及「臨時建築物」移出「工地」。

(3) 本條的任何規定皆不會損害「承建商」行使他有權得到的任何其他權利或補償，及所有代替或增加本條所指定的權利和補償。

Commentary

This Clause is seldom invoked because the Government of the HKSAR is not in a position or unlikely to be incapable of making timely payment to the Contractor. Counting the time, under Clause 79(1), the Engineer shall certify within 21 days and then payment shall become due in another 21 days. Then any non-payment shall be subject to interest under Clause 79(4). After another 28 days, the Contractor may give 14 days' notice in writing to the Employer to make payment of the sum due, referring to this Clause. If the Employer does not make payment within the 14 days of notice, the Contractor can terminate the Contract. The Constructional Plant and temporary buildings brought onto the Site by the Contractor vested in the Employer shall now be re-vested in the Contractor who shall remove those from the Site with reasonable despatch.

Analysis and application

Sub-clause (1) is straightforward and has been described in the commentary. Sub-clause (2) merely states that the property of Construction Plant and temporary buildings shall be re-vested in the Contractor who should remove those from the Site as soon as reasonable. Sub-clause (3) states that the right of the Contractor against the Employer's breach of Contract is not limited to the provision of this Clause.

Comparison with the ICE Conditions

This ICE Clause 64 mainly provides for the Contractor to terminate his employment in the event that the Employer becomes bankrupt or being insolvent which affects his ability to make payment to the Contractor. The termination only requires 7 days' notice but the Contractor can extend the period to give the Employer an opportunity to remedy the situation. The English law in any case allows the Contractor to suspend work for failure of the Employer to make payment.[1]

Comparison with the FIDIC Conditions

The FIDIC Clause 16.2 lists several conditions regarding Employer's failure to make payment or show necessary financial arrangements, or perform his other obligations, or become bankrupt or insolvent. Any of these shall entitle the Contractor to terminate the Contract either within 14 days' notice or immediately under the conditions in sub-paragraphs (f) and (g).

Notes

1. See Roger Knowles, *150 Contractual Problems and Their Solutions*, Section 7.6, and the *Housing Grants, Construction and Regeneration Act* 1996, Section 112.

Suggested Answers to Questions

Question 1

Since the TCSS contractor is employed by the Government, he is called Specialist Contractor, which is defined as "any contractor employed by the Employer to execute Specialist Works." Under Clause 20, the main Contractor is not responsible for the safety of the operation of the Specialist Works. However, under Clause 21, the main Contractor is responsible for the care of the Specialist Works.

Question 2

Though the back-span is supported by "specially designed falsework," it is part of the Works and the contractor carrying out such work is domestic sub-contractor. Under Clause 20, the main Contractor is responsible for the safety of its operation and under Clause 21, the main Contractor is responsible for the care of such works.

Question 3

Loosening the ground for hydroseeding is the responsibility of the Contractor and the Engineer needs not give the order to do so. Even if the Engineer had given the order, it does not entitle the Contractor to any other payment except payment under the original terms of the Contract, since Clause 2(4) states that "no act or omission by the Engineer ... shall in any way operate to relieve the Contractor of any duties, responsibilities ..."

Question 4

There is a discrepancy between the Specification and the Drawing. The Contractor should write to the Engineer asking for clarification and according to Clause 5, the Engineer shall issue instruction to clarify such discrepancy. According to sub-clause (b), if the compliance with such instruction shall involve the Contractor in expense which the Contractor did not or had no reason to anticipate, the Engineer shall value such expense. In the case of the spacing of the bolts which needs to have 75mm clear width, the Contractor needs to rectify and it is likely that the Contractor should be compensated.

Question 5

This depends on the design requirements stipulated in the Contract, usually found in the Specification or the Drawings. This problem of matching levels is not only limited to noise barriers but also to parapets and surfacing. Engineers experienced in bridge construction should be able to write into the Contract a procedure or design requirement which the experienced Contractor can follow to complete the work, e.g., calculation of precamber, etc. In the unfortunate event that the noise barrier components do not fit into the as-built bridge structure, the determination of liability depends on how the Contract is written and what subsequent actions have been taken by the Engineer and the Contractor, i.e., whether clarifications have been requested by the Contractor or the instructions given by the Engineer.

Question 6

The Drawings for the specially designed falsework are supplied by the Contractor and therefore in accordance with Clause 8(2), the Employer may use any information provided by the Contractor but he shall not divulge such information except for the purpose of the Contract or for the purpose of carrying out any repair, amendment, extension or other work connected with the Works. In this case, the Employer wants to extend the back-span and this falls within the provision of Clause 8(2).

Question 7

This is a typical case illustrating the need to inspect the Site and the Contractor has to satisfy himself, before submitting his Tender, as regards existing roads or other means of communication and access to the Site under Clause 14. A weekday inspection reveals the traffic queue and therefore the Contractor claiming remedy must fail.

Question 8

This question involves the principle of risk-sharing under the Contract. Under Clause 13, the Contractor "shall be deemed to have examined and inspected the Site and its surroundings and to have satisfied himself, before submitting his Tender, as regards existing roads ... access to the Site, the nature of the ground and subsoil ... and generally to have obtained his own information on all matters affecting his Tender." Under Clause 14, the Contractor "shall be deemed to have satisfied himself before submitting his Tender as to the correctness and sufficiency of his Tender." Therefore, the Contractor has to take the risks arising from traffic, utilities and environmental constraints. However, regarding underground utilities, if uncharted utilities are found which critically delay the Works, the Contractor may claim for extension of time under Special Circumstance or any appropriate SCC Clauses regarding utilities obstruction. In the more difficult cases where the utilities, whether charted or uncharted, create a situation where their diversion is impossible or impracticable within reasonable time frame, the Contractor may be relieved of his responsibility for the concerned work under Clause 15. Under such case, the Engineer needs to issue variation order to avert such impossibility or impracticability. In the case where the anticipated time for diversion of utilities are very substantial and may be less desirable for the project, the Engineer may also issue variation orders to assist the progress of the Works. Under such circumstances, the Engineer may take into account the original reasonable time for diverting such utilities in considering whether an extension of time should be granted to the Contractor. To a lesser extent, the above cases also apply to environmental and traffic constraints.

Question 9

The Clause 16 programme shows the Contractor's intending sequence, method and timing for carrying out the Works and his working arrangement, Constructional Plant and Temporary Works. The programme also shows the various critical paths to the completion of the Works. The Engineer needs to grant extension of time for the completion of the Works and pay prolongation cost under Clauses 50 and 63 respectively only if the delay impacts on the critical paths of the approved programme. Even though the Employer's delay impacts on the critical path, the Contractor may not be automatically entitled to extension of time because he has a general duty to mitigate the delay using reasonable endeavour (see Clause 50, Note 8). Apart from that, the Mega Project Conditions of the HKSAR impose a duty on the Contractor to submit a revised programme and the mitigation measures the Contractor proposes to adopt. The Engineer can take the Contractor's action or non-action into account when considering any extension of time.

Question 10

To distinguish Temporary Works and Constructional Plant, we look at the definition under Clause 1. "Temporary Works" means all temporary work of every kind required for the construction, completion and maintenance of the Works while "Constructional Plant" means all appliances or things of whatsoever nature required for the execution of the Works but does not include materials or other things intended to form or forming part of the permanent work or vehicles in transporting any personnel, Constructional Plant, materials or other things to or from the Site. We therefore see that the definition of Constructional Plant is much wider, including also the Temporary Works. Therefore, all the trusses and specially made lifting appliances at the column head, and even the temporary concrete columns are Constructional Plant. The fixed trusses and temporary concrete columns are Temporary Works.

Question 11

Under Clause 20(2), the Contractor shall in connection with the Works provide and maintain all lights, guards, fences and warning signs. The provision of guarding, lighting to the public road and fencing to the private development are in connection with the Works and therefore the Contractor has to pay for it. Under Clause 13 regarding a deeming provision for the inspection of the Site before submitting the Tender and Clause 14 concerning sufficiency of the Tender, the Contractor is not entitled to claim even he has provided more lighting, guarding and fencing than he envisaged at the Tender. However, under the ICE Conditions Clause 11, "The Contractor shall be deemed to have inspected and examined the Site ... and to have satisfied himself so far as is practicable and reasonable before submitting his tender ..." And together with its Clause 12 regarding compensation for adverse physical condition, the Contractor may be entitled to claim extra if the lighting, guarding and fencing he provided are more than he could reasonably envisaged. Similar provisions are found in the FIDIC Conditions.

Question 12

Under Clause 53(5)(b), "The Engineer, following a written request from the Contractor, may give a certificate of completion in respect of any substantial part of the Works which has been completed to the satisfaction of the Engineer … and is capable of permanent occupation and/or permanent use by the Employer." The completion of the cable-stayed bridge tower, though being very substantial and expensive, does not satisfy the above condition of permanent occupation or use and therefore the partial completion certificate cannot be granted.

Question 13

Under Clause 22(2), "The Contractor shall make good or at the option of the Employer shall pay to the Employer the cost of making good any damage … to any property of the Employer …" The Sewage Treatment Work is the property of the Government and therefore the Contractor is responsible for any accidental damage.

Question 14

Regarding insurance, any component of the cable-stayed bridge is very expensive and it is desirable to buy Works insurance covering all the Works. The Site is in the proximity of the public road and the container operation and therefore third party insurance is inevitable. Under the law, the Contractor has to buy employees' compensation insurance for all his employees, and especially the workmen. It should include the workmen of all his sub-contractors if separate insurance policies are not taken out by the sub-contractors. The position with the self-employed is more complicated and generally they should provide their own insurance policies. If there is any design by the Contractor, the Contractor, his designer and independent checker should take out professional indemnity insurance. In most cases, the Contractor also insures his plant.

Question 15

The container boxes being blown off causing damage to the Contractor's falsework system is a damage to the Contractor's Plant or Temporary Works, or subsequent damage to the permanent work of the Employer. The Contractor may sue the container operator under the law of tort, and especially the container operator's negligence. In reality, such losses are covered by insurance policies and the insurance company will pay but recover from the container operator. If there is any excess to be paid by the Contractor himself, he can recover the excess from the container operator.

Question 16

The tower crane is the Plant or Temporary Works designed and constructed by the Contractor who is fully responsible for its safety. If it falls onto the adjacent container operator's land, under Clause 22(1), the Contractor shall indemnify the Employer against all claims for damage by the container operator and he must be responsible for all the consequences. Usually the Contractor takes out third party insurance and except for paying the excess, he will be covered by this policy.

Question 17

The lighting term contractor is employed by the Employer and is a Specialist Contractor. The erection of the lighting post is not part of the Works but is Specialist Works as defined in Clause 1. Under Clause 20, the Employer is responsible for the safety of its own Specialist Works. Therefore the Contractor will not be responsible for any damage to property arising therefrom. Therefore, the damage to the private car is the Employer's own responsibility. As regards the damage to the adjacent columns, since the Contractor is not responsible for the safety and stability of the lighting post, the damage to the columns will also be the Employer's own responsibility.

Question 18

Under Clause 30, the Contractor has to comply with the law and addition and amendment to the law, and therefore in the absence of any Special Conditions, the Contractor has to engage the service of the two structural engineers at his own expense. However, as described in the general commentary to Clause 30, a Special Condition allows compensation to the Contractor for any change in law described in the Schedule attached to the Special Condition. In this case, the amendments are most likely to be made under the *Builder's Lift and Tower Working Platform (Safety) Ordinance*, which is described in the Schedule and therefore the Contractor is entitled to compensation.

Question 19

This is a high-level construction Site of major safety concern and hence the *Factories and Industrial Undertakings Ordinance* must be observed. Since lifts transporting workers and people are provided, the *Builder's Lift and Tower Working Platforms (Safety) Ordinance* is applicable. The construction noise must be abated to the standard required by the *Noise Control Ordinance*. The Site also involves waste disposal and hence the *Waste Disposal Ordinance* must be observed. The workers on Site must be subject to the *Employment Ordinance*, and also the *Immigration Ordinance* if foreign workers are involved. The *Employees' Compensation Ordinance* governs the compensation to workers in case of any injury. The electricity and gas operations must be subject to the *Electricity Ordinance* and *Gas Safety Ordinance*. Finally, all construction activities involving public roads must comply with the *Road Traffic Ordinance*.

Question 20

The erection of gantry before the strength of structural steel is tested is at the Contractor's own risk. Under Clause 46, the Engineer can order the Contractor to remove unsatisfactory works. If removal of the completed works is not practical, the Engineer can order a variation of the works in lieu of removal at no extra cost to the Employer. In this case, however, there is no need for the removal or a variation. The Engineer can simply accept the slightly substandard material and determine a reduction in price under a Special Condition to GCC Clause 46.

Question 21

The decision not to divert the overhead cable appears to have come from the utilities company, which is a third Party to the Contract. However, since the Employer have incorporated into the Contract that the cable will be diverted, the subsequent decision, whoever makes it, may be regarded as a change of contract requirements in accordance with Clause 60, which empowers the Engineer to order any variation necessary or desirable for the completion of the Works. The accommodation of the overhead cable involves a change of method and timing of construction. Therefore the Contractor can claim extension of time in accordance with Clause 50 and disruption costs in accordance with Clause 63.

Question 22

The Engineer approved the Contractor to build the gantry at the location and the gantry was subsequently built. To impose the restriction of loading activities to non-peak hours by the Marine Department is similar to restricting loading and unloading in a stretch of road adjacent to a construction site by the Transport Department. The Marine Department is in fact carrying out its statutory function. The Contractor may claim extension of time under Clause 50 "Special Circumstance", but he will not be entitled to claim costs.

Question 23

If the lack of progress is due to the fault of the Contractor, the Engineer is "empowered to instruct the Contractor in writing to carry out the Works, or any part thereof during any hours of the day" And he may therefore instruct the Contractor to carry out night work. However, before giving such instruction, he must be sure that all likely extension of time has been granted to the Contractor or otherwise the Contractor can claim extra payment in accelerating the Works.

Question 24

Clause 51 allows the Engineer to require the Contractor to take steps to expedite the completion of the Works, including carrying out the Works during any hours of the day. Instructing the Contractor to provide more plant specifically designed by the Contractor is rather unusual and it may not be a valid instruction. If it has to be done, it would be prudent to sign a Supplementary Agreement with the Contractor, specifying each Party's responsibility. The design and the operation of the Specific Plant must be made the responsibility of the Contractor.

Question 25

The achievement of a "Stage" is different from the completion of a Section in that the responsibility for the care of the part of Works achieved under a Stage is not passed to the Employer 28 days after the achievement of the Stage, but in the latter, the responsibility for the care of Works is passed to the Employer. In order the allow the container vehicles to pass between the columns, we have to ensure that the responsibility for the care of Works still lies with the Contractor and therefore a staged completion of the columns has to be specified. Apart from that, it has to be stated in the Contract that the container vehicles are allowed to pass between the columns.

Question 26

Under Clause 53(5)(b), the Engineer, following a written request from the Contractor, may give a certificate of completion in respect of any substantial part of the Works which has been completed to the satisfaction of the Engineer and is capable of permanent occupation and/or permanent use by the Employer. Regarding this particular case, the completion of one viaduct of the dual 3-lane road is undoubtedly a substantial part of the Works. Also, the phrase "capable of permanent occupation and/or permanent use by the Employer" does not require actual occupation or use, and even though the Employer does not use it, the Engineer can still certify partial completion. When certifying partial completion, the Engineer should ensure that any outstanding works are completed within the Maintenance Period of that particular part, which starts to run upon certification of that part.

Question 27

Upon completion of the flyover ready for opening to traffic, the Engineer can have two choices. He can either certify the completion of the whole of the Works leaving the ground level works as outstanding Works, or he can certify partial completion of the flyover. If he adopts the second approach, he can still have control in the completion of the ground level works. If he adopts the first approach, he has to make sure that the majority of the ground level works are completed within the few months after certification of completion because the Maintenance Period for the ground level works is effectively reduced.

Question 28

It is obviously better to investigate the cause of the movement before covering up the slope with permanent surface protection and therefore the Engineer should first suspend part of the slope works under Clause 54(1). Under sub-clause (2)(d), the suspension is for the safety of the Works and the safety of the person and therefore the Contractor may not be entitled to any time extension nor cost reimbursement. Then the Engineer may instruct the Contractor to investigate the cause of the slope movement under Clause 58. Depending on the results of investigation, if the Contractor is responsible for the defects, he has to propose remedial works under Clause 46, generally at no additional time and expenditure to the Employer. However, if the Employer is at fault, i.e., design fault, then extension of time and reimbursement of expenditure should be granted to the Contractor under Clause 58(3) and/or the exception to Clause 54(2)(d). Obviously, the Engineer has to minimize the time for investigation and require the Contractor to implement temporary protective measures to the concerned slope.

Question 29

The Labour Department is a Government Department but it is acting in its statutory capacity to suspend the Works. The sliding of the girder is the fault of the Contractor and therefore he should be responsible for its own delay and costs. The Engineer should just carry out his own investigation and request the Contractor to submit any preventive and improvement proposals. He could also request the Contractor to submit a revised programme to ensure timely completion of the Works and to take steps as necessary to expedite the completion of the Works under Clause 51(1). As regards the injures of the workmen, the Contractor shall be responsible and it would be compensated under the employees' compensation insurance, which is mandatory under the law.

Question 30

The need to install a new automatic smoke detection system after the issue of a Certificate of Completion requires the Engineer to order such work during the Maintenance Period. Though the HKSAR General Conditions of Contract does not allow for ordering variation during the Maintenance Period, a Special Condition has been incorporated into all HKSAR Contracts, which allow the Engineer to do so (see the comparison of Clause 60 with the ICE Conditions on p. 147).

Question 31

When considering whether substantial completion of the Works can be granted, the basic test is the readiness for "operational or functional occupation" (see p. 116). Out of the four kilometers of twin decks, only two sections of parapets are outstanding. By suitable temporary traffic measures, the bridge decks can be opened to traffic and hence there is no difficulty in certifying substantial completion.

Question 32

When considering the measurement of the Works, we must first refer to the Standard Method of Measurement (SMM) because under Clause 59(1), the Bill of Quantities shall be deemed to have been prepared in accordance with the procedures set forth in the Method of Measurement. According to the SMM Clause 8.30–8.32, concrete profile barriers shall be measured along the centre-line of the profile barriers in linear metres. The item coverage for concrete profile barriers includes precasting, transporting and fixing in position, jointing, ducts and drain pipes, filling and compaction to core, concrete topping, etc. The concrete barriers in the photo consist of a wide earth-filling but no concrete topping, possibly for planting purposes. A particular preamble has to be written to cover the situation. A typical section showing the centre-line of the profile barrier has to be shown.

Question 33

This depends on whether the median drainage has been particularized in the Specification and/or in the Bill of Quantities, including its location. Under Clause 5(2), the Engineer shall issue to the Contractor instructions clarifying the ambiguities and discrepancies. If the compliance of instruction shall involve any expense by reason of ambiguities or discrepancies the Contractor did not and had no reason to anticipate, the Engineer shall value such expense. Also, if extra time is involved, the Engineer can grant extension of time in accordance with Clause 50(1)(a)(iii).

Question 34

The model may sometimes be useful in arbitration on the valuation of cost because it can be used together with photographs to give a three dimensional appreciation of the Site conditions. The model can give the arbitrator a good appreciation on whether any works are carried out under similar conditions and circumstances to any item of work priced in the Contract, in accordance with Clause 61(1)(b) and (c).

Question 35

In making such valuation, the Engineer should first base on the rate of the 1.2m high parapet. He should break down the work into various components, i.e., formworks, concrete, reinforcement, etc., and assess the costs of these components. From these costs, the Engineer then builds up the Cost for the 1.5m high profile barrier.

Question 36

Though under Clause 62 the Engineer can order work on a daywork basis, it may not be appropriate to do so if the works require intense planning and with some design element. The FIDIC Conditions Clause 13.6 specifically says that daywork is for works of a minor or incidental nature. The HKSAR Conditions Clause 62 does not expressly say so, but there are internal guidelines for ordering daywork. For the electrical components missing from the Drawings, warranties and some design are usually required and ordering daywork may not be appropriate.

Question 37

For the removal of the dangerous boulder on the slope, Temporary Works design is inevitably required and ordering daywork may not be appropriate. A proper variation order should be issued which should include design and checking. If the boulder falls down, the Contractor has the primary responsibility and the designer and checker would also be liable. If, however, daywork is ordered, the Contractor may not be fully responsible and the Engineer may have some liability depending on the manner he orders the daywork.

Question 38

If the precast segments have to be cast 200 mm shorter than the original design, the Engineer has to issue a variation order under Clause 60. Under Clause 63(c), the Contractor may, by reason of the progress of the Works or any part thereof having been materially affected by the variation order, seek reimbursement. The direct cost of the variation order, such as extra construction joints, formworks, etc., may be evaluated under the variation order, while any delay or disruption to the progress of the Works can be separately valued under Clause 63(b).

Question 39

For the laying of additional water-cooling main, the Contractor should first request the Engineer to issue a variation order. The Contractor should keep all contemporary records, such as payment vouchers, labour, Plant and materials used including all idling time, for the valuation of the variation order under Clause 61 by the Engineer. It is likely that the Engineer's variation order would cover the direct costs. For any indirect costs, such as disruption to progress of work and prolongation of time for completion, the Contractor has to submit a notice of claim under Clause 64(2) within 28 days after the Engineer's instruction. Under Clause 64(3), the Contractor has to keep contemporary records and agree on such records with the Engineer's Representative. Under Clause 64(4), the Contractor has to submit interim accounts with full and detailed particulars supporting his claim and any further accounts as necessary.

Question 40

During examination times, the Contractor is required to mitigate the noise, and he may not be able to use powered mechanical Plant to carry out the work, or he may have to erect temporary noise screens to reduce the noise. This involves a change of method of working and/or additional Temporary Works and will be effected by the Engineer issuing a variation order to the Contractor. The Contractor should furnish relevant information to enable the Engineer to value the variation order. If the Contractor wishes to submit a claim for disruption to the progress of work and overall delay in the time for completion, he should follow the procedures stipulated under Clause 64.

Question 41

While the Engineer reviews the details of reinforcement in the cross-beams, the launching girder will be idling and other works for the deck will be suspended. The Contractor should:
(1) request the Engineer to suspend that part of the Works under Clause 54(1);
(2) apply to the Engineer who shall ascertain the Cost incurred under Clause 54(2) and the Engineer shall then certify the Cost under Clause 79.
If the Engineer refuses to issue a suspension order, the Contractor can submit a notice of claim under Clause 64(2) and then follow the procedures in Clauses 64(3) and (4).

Question 42

According to Clause 1(1), the Provisional Sum means a sum provided for work or expenditure which has not been quantified or detailed at the time the Tender documents are issued. It may have a brief description in the Bill of Quantities. If such a sum has been provided for the video work, the Engineer can order such work and value in accordance with Clause 61, otherwise the Engineer may order such work through a variation order under Clause 60 using the Contingency Sum

Question 43

According to Clause 81(1), if the Contractor gets bankrupt, the Employer can enter upon the Site and may use the Constructional Plant, temporary buildings and materials which become the property of the Employer under Clauses 71 "Vesting of Constructional Plant and Temporary Buildings" and 72 "Vesting of Materials". According to Clause 1, Constructional Plant includes the falsework system and the crawler crane. By the same Clause 81(1), the Employer may sell any of the Constructional Plant, temporary buildings and unused materials for the satisfaction of any sum due to the Employer under the Contract.

Question 44

Under the HKSAR Conditions Clause 79(1)(c), the Contractor may get interim payment for the value of materials for inclusion in the permanent work not being prematurely delivered to the Site. Since the steel deck unit is in Mainland China, the Contractor cannot claim interim payment unless any Special Conditions provide for that. Even so, a set of procedures has to be complied with and usually an offshore bond is required from the Contractor. If the ICE Conditions were used, under its Clause 60(1)(c), the value of a list of any of those goods or materials identified in the Appendix to the Form of Tender, which have not yet been delivered to the Site but the property of which has seen vested in the Employer pursuant to Clause 54 can be included in the interim payment.

Question 45

Under the HKSAR Condition Clause 79(1)(b), the Contractor can get interim payment for the value of Temporary Works for which a separate sum is provided in the Bill of Quantities. If the launching girder is not included in the Bill of Quantities, no interim payment can be made. Also, the launching girder is arguably not Temporary Work but is Constructional Plant. For the ICE Conditions, under Clause 60(1)(d), the value of the Contractor's equipment for which separate amounts are included in the Bill of Quantities can be included in the interim payment. Under the FIDIC Conditions, interim payment can be made to the Plant intended for the Works under Clause 14.3(e).

Question 46

Under Clauses 79 and 52, the Employer can deduct liquidated damages from the interim payment. However, if the Contractor's cash flow to complete the Works is seriously affected, it is not uncommon that the Engineer, after taking account of the overall situation, recommends to the Employer to withhold the deduction of liquidated damages. This is especially the case if there are claims for extension of time and/or money, which have not yet been finalized. Also, if there is a bond which can cover future deduction or if the retention money is adequate, then the Employer can withhold the deduction.

Question 47

Under Clause 58, prior to the issue of the maintenance certificate, the Engineer can instruct the Contractor to investigate the cause of any defects. If the defects are the ones which the Contractor should be liable in accordance with the Contract, the Contractor shall bear the cost of investigation and subsequent making good of the defects. If, however, the defects are discovered after the maintenance certificate is issued, the Contractor shall only be responsible if such defects could not with reasonable diligence have been discovered by the Engineer. Similar to the HKSAR Conditions, latent defects discovered after the issue of the Defects Correction Certificate (Clause 61(2)) under the ICE Conditions or the Performance Certificate (Clause 11.9) under the FIDIC Conditions are still the liability of the Contractor.

Question 48

According to Clause 81(1), the Employer can sell the Constructional Plant and apply the proceeds of sale towards the satisfaction of any debt owed by the Contractor to the Employer. Since $10M is still owed to the Employer and the Plant is worth $8M, the Employer can sell the plant and retain all the money.

Question 49

If the Contractor goes bankrupt, the Employer may follow the following steps under Clause 81:
(a) He shall give at least 7 days' notice to the Contractor and then enter upon the Site and expel the Contractor from the Site.
(b) He shall make a record of the state of completion of the Works and update all measurements; he shall also make a complete record of the Constructional Plant and materials on Site.
(c) At the same time, he shall decide the approach to complete the Works by using his own workforce or call a new contract; he can make use of the Constructional Plant and materials which have been vested upon him under Clauses 71 and 72.

(d) If the Employer decides to retain the service of the existing sub-contractors or suppliers, he can require the Contractor to assign the benefit of any sub-contracts to him.

(e) Under a Special Condition, the Engineer can certify money due to either party when the difference between the following can be ascertained:

   (i) such sum as would have been due to the Contractor if he had completed the Works together with any proceeds of sale; and

   (ii) the Cost of completing the Works, with damages for delay and other expenses properly incurred by the Employer.

Question 50

Although the outstanding works are outside the Site limit, provided those works are included into the Contract, the Contractor has no right to refuse carrying out those works. However, the Engineer must ensure that the land is available from the Lands Department and check that the insurance covers those works. If the Contractor still refuses to carry out the works, the Engineer may give the Contractor 14 days' notice in writing to carry out such works under Clause 82(1). If the Contractor still fails to do so, the Employer shall be entitled to carry out such works by his own workers, e.g., his maintenance contractor. All additional expenditure properly incurred by the Employer shall then be deducted from the Contractor's account under Clause 83.

Question 51

This situation will put the Employer into a very difficult situation and it is very undesirable even though the Employer has the power to carry out such work by invoking Clause 82 for several reasons:

(a) Though the Contractor's claims for EOT and money are apparently unjustified from the Engineer's point of view, the decisions of the Engineer are subject to review by the future arbitrator, and if the arbitrator rules in favour of the Contractor, the Employer may have to compensate the Contractor for any disruption, prolongation and loss of profit.

(b) The works to be completed by the new Contractor may have technical interface with the Contractor's completed works and there may be uncertainty as to each Party's responsibility in case of any future failure of the work.

(c) It may be difficult for the Employer to find a new Contractor to complete the Works. The maintenance contractor may not be technically competent enough to carry out such work or unwilling to do so because of the low Contract rates. Calling a new contractor will take at least several months.

(d) The Contractor may insist that he has a right to complete the work himself even if he risks himself exposing the liquidated damages being deducted by the Employer.

Having considered all the above, the Employer may first use all administrative means to oblige the Contractor to co-operate. Such means include the consideration of giving adverse performance reports and suspending the Contractor from tendering future works. The Employer, after the decisions from the Engineer, can propose adjudication as a quick means to decide on the claims. As a last resort, the Employer only then uses his own workers or maintenance contractor to complete the works to meet the deadline. See Note 1 of Clause 82.

Question 52

The Contractor can raise a dispute anytime and ask for the Engineer's decision under Clause 86. If the dispute cannot be resolved by the Engineer's decision, the Contractor can propose mediation or adjudication (under a Special Condition). The Employer is likely agreeable to mediation or adjudication. However, the Contractor cannot commence arbitration until the alleged completion of the Works under Clause 86(3), unless agreed by both Parties.

Question 53

The *Arbitration Ordinance* (Cap. 341) allows the parties to make an agreement in their contract to submit their dispute to arbitration, but that does not mean that the Court can be put aside altogether. When one party challenges the jurisdiction of the arbitrator to decide on the particular dispute, the arbitrator can submit to the High Court for a ruling. The arbitrator can also submit a preliminary question of law in connection with the dispute to the Court. When the arbitrator has made an award, the parties can have some limited rights of appeal to the Court. Finally, the Court is involved in enforcing the arbitral award.

Table of Cases and Ordinances Quoted

Index of Clauses / sub-clauses

Bibliography

Abrahamson, Max. *Engineering Law and the ICE Conditions*, 4th Edition. London: Applied Science Publishers, 1979.

Booen, Peter L. (Principle drafter). *The FIDIC Contracts Guide*. Geneva: FIDIC, 2000.

Downes, T. Antony. *A Textbook on Contract*, 4th Edition. London: Blackstone Press, 1995.

Eggleston, Brian. *A User's Guide to the ICE Conditions of Contract*, 7th Edition. Oxford: Blackwell, 2001.

Government Practice Note on GCC61 and 63(b) (HKSAR).

Hawker, Geoffrey. *ICE Conditions, 6th and 7th editions Compared (Measurement Version)*. London: Thomas Telford, 1999.

Housing Grants, Construction and Regeneration Act 1996 (UK).

ICE Conditions 7th editions (Measurement Version) Guidance Notes. London: Thomas Telford, 1999.

Jones, Michael A. *Textbook on Torts*. London: Blackstone Press, 1995.

Knowles, Roger. *150 Contractual Problems and Their Solutions*. Oxford: Blackwell, 2005.

Ramsey, Vivian and Furst, Stephen. *Keating on Building Contracts*, 6th Edition. London: Sweet & Maxwell, 2003.

Recommendations in Grove Report on General Conditions of Contract (1998). (HKSAR Government).

Wallace, Ian Duncan. *A Commentary on the FIDIC International Standard Form of Engineering and Building Contract*. London: Sweet & Maxwell, 1974.